The Bearded Lady Project

The Bearded Lady Project
Challenging the Face of Science

Edited by Lexi Jamieson Marsh and Ellen Currano

Photographs by Kelsey Vance and Draper White

COLUMBIA UNIVERSITY PRESS
NEW YORK

Columbia University Press

Publishers Since 1893

New York   Chichester, West Sussex

cup.columbia.edu

Cataloging-in-Publication Data available from the Library of Congress

LCCN 2019954454

ISBN 978-0-231-19804-2 (cloth)

ISBN 978-0-231-55246-2 (electronic)

Columbia University Press books are printed on permanent and durable acid-free paper.

Printed in the United States of America

Managing Editor: Matt Harrison

Cover image copyright Kelsey Vance

Book design: Patrick Côté

Photograph Credits

Kelsey Vance: ii and iii, vi and vii, viii, xii, xvi, xx and xxi, xxii, 2, 5, 6, 14, 20, 23, 24, 25 and 26, 32 and 33, 34, 35, 36, 38 and 39, 40, 44, 48, 52, 53, 54, 59, 60, 61, 65, 70, 73, 74, 75, 76, 77, 80, 86, 89, 90, 91, 92, 93, 94, 96, 98, 99, 100, 104, 107, 108, 109, 110, 111, 112, 113, 114, 115, 122, 124, 126, 127, 128, 129, 133, 135, 136, 137, 138, 139, 140, 141, 142, 144, 146, 148, 150, 151, 153, 156, 162 and 163, 164 (left top two), 165 (bottom), 166, 167, 168

Draper White: iv and v, 46, 56, 57, 66, 69, 88, 158, 164 (left bottom), 165 (top), 169, 170, 171 (bottom), 173, 183, 184 and 185

Lexi Jamieson Marsh: 160, 171 (top)

Laura Dempsey: 62

Ellen Currano: 17

This book is dedicated to women in science struggling to be seen and heard.

TEXTS

PORTRAITS

Foreword

Lexi Jamieson Marsh

The Bearded Lady Project started as a wry joke. It happened five years ago during a dark and intimate dinner conversation with my friend and personal hero, the accomplished plant paleontologist (paleobotanist) Dr. Ellen Currano. We both were near despair over our struggles to be accepted into professional worlds dominated by men. Ellen sought to lighten the mood with a funny suggestion that would seemingly right all of our wrongs: "Maybe if I just put a beard on my face. . . ."

The image struck a chord in that moment—that all Ellen would need in order to do her job would be to look like everyone else. Would simply adding a beard validate her existence? It was laughable that this quick fix could potentially right generations of wrongs toward women in the workplace.

Meeting Ellen at the restaurant in my usual flannel shirt, jeans, sneakers, and baseball cap, I was feeling pretty low about my own future in filmmaking. I was exhausted from shooting a challenging commercial with a film crew in Cincinnati—not because of long hours or technical difficulties, but because of the defeating realization that my all-male colleagues found me distracting, problematic, and out of place on set. They snickered over what I would look like if I ever "cleaned up." On that same set, a male superior informed me that women "always say they want to direct, but when given the opportunity, they realize it's just not for them."

Before meeting Ellen, I figured that all my dinosaur-inspired dreams of becoming a real-life paleontologist evaporated around age seven, along with aspirations of walking on the moon or getting elected president. I still couldn't believe Ellen traveled the world to discover ancient fossils for a living. Not only was she a paleontologist, but she was an exceptional one. Ellen had worked at the Smithsonian and, before finishing her postdoc or celebrating her thirtieth birthday, had landed a tenure-track job at Miami University in Oxford, Ohio—the location of that night's good-bye dinner,

Lexi Jamieson Marsh films Ellen Currano at work in the Hanna Basin, Wyoming.

as she had already been recruited to a new position at the University of Wyoming.

Ellen had been an inspiring friend over the past five years. Every time she returned from one of her months-long field trips to Ethiopia or Wyoming, I eagerly awaited the demise of her jet lag until she was ready to talk about her recent adventure. A close encounter with rattlesnakes during a thunderstorm; battling a baboon for a bag of oranges—hers were tales a superhero would tell. When I shared my difficult day on set with her, she admitted that she was treated the same way as a paleontologist. Her revelation broke my heart. The tears, impossible to hold back, blurred Ellen and the restaurant. The trials of my day still stung, but the reality that one of the best and brightest individuals I knew was being treated poorly among her peers broke me. *If a superstar like Ellen can't be respected for the work she does*, I thought, *how can an unknown filmmaker like me expect anything different?*

Steering away from the path of self-pity, Ellen turned our night around: "I mean, maybe if I could put a beard on, then I could actually do my job."

That was the origin of this project. I couldn't forget about the absurd idea that Ellen's workplace problems might be solved not by all of her hard work and labor, but simply by putting on a beard. At two in the morning, I got out of bed, opened my laptop, and wrote to her: "Would you seriously wear a beard?" In retrospect, I realize that in a crucial and defeating moment, I was more willing to come to the rescue of a friend than I was to stand up for myself.

I attribute the ongoing success of *The Bearded Lady Project* to a handful of women who early adopted our unconventional means of challenging gender stereotypes. The original idea was to make a five-minute video to be shared on YouTube. But thanks to the many women and institutions who found value in our creation, our idea grew into two documentary films, a portrait exhibition, a scholarship fund, and now a book. But if I go all the way back, before anyone was recruited, before beards were purchased and our cameras started rolling, first and foremost the reason we are here today is my mother, Ljuba Marsh. She was the first person with whom I shared our nutty and embarrassing idea. It is always an out-of-body, surreal, and heightened emotional moment when you put your imagined concept to words. The idea was unnamed, unrefined, but one that I couldn't shake. As an artist, being able to express your ideas, especially at an early conceptual stage, is daunting. But my mother, my lifelong advocate, had always been there to listen to my ideas and would support them no matter how unpolished they were.

At this time, I was working with a very broad concept. I knew nothing of the difference between paleontologists and geologists, but I knew the message. I knew I wanted to be a voice for my friend, and to use film in a way that would have some positive impact in her life and hopefully, a positive influence on others.

The pitch to my mother was long-winded and fairly incoherent, but her response was direct and, from my perspective, off topic: "What do you need to do to get this started?" I didn't know how to respond. She rephrased: "What would be the very first practical step you would need to take to begin filming?" An endless amount of physical and financial hurdles stood in the way of giving her an appropriate response. After a few attempts to answer, I boiled it all down to filming Ellen in the field. "So really, all you need is a plane ticket," she responded. She was right, but as simple and as straightforward as that was, I was still unable to convince myself that now was the right time to start.

The next morning my mom placed a check on the counter where I was sitting, having coffee. "This will get you to Wyoming," she said with a grin. "There are no strings attached, but I do want you to promise me that you won't back down." I took that flight to Wyoming. From April to July 2014, we officially started our adventure.

The Bearded Lady Project is about challenging the negative stereotypes in science, but its foundations are friendship, love, and support. It thrives on women supporting women: on a mother believing in her daughter and a friend fighting for the world to see her friend in the same way she always has.

This book, like this project, is a collaboration between art and science. It is a mix of scientific studies, personal essays from scientists who participated, and a glimpse of how two women made this project from start to finish. It takes a filmmaker to fight for a paleontologist. It takes a paleontologist to risk everything in support of a filmmaker. This project is stronger because of the variety of voices; it is interdisciplinary collaboration at its finest. I hope you will take away the lesson that I have learned: every one of us has the power to connect, to challenge, to advocate, and to remind each other that we are individually inspiring and collectively empowering.

PART I
Why Challenge the Face of Science

Marieke Dechesne

Geologist, United States Geological Survey

1

"Pictures in Our Heads": Challenging Stereotypes of Scientists and Science

Amanda Diekman

In 1922, the American journalist Walter Lippmann coined the term "stereotype" to describe how we each hold "pictures in our heads" of certain groups or situations, and how these preexisting beliefs profoundly influence our reactions.

Nearly 100 years later, social psychology has amassed a nuanced understanding of how stereotypes work, when they are most likely to influence decisions and behaviors, and what we can do to reduce their detrimental impact. Stereotypes are powerful, but we are not powerless before them.

Stereotypes are cognitive shortcuts that allow quick judgments in a busy and demanding world. The problem, of course, is that valuable information is lost in this simplification. When we know that our judgments might be influenced by stereotypes, we can slow down, reconsider information, and think about people as individuals rather than in categories. The first step, though, is to become aware of how these expectations might play into our decisions.

The Bearded Lady Project challenges conscious and unconscious expectations about who paleontologists are, what fieldwork entails, and what science is. These portraits highlight how difficult it can be to fit to implicit and explicit demands of a particular role or set of expectations. The beard on a field site is a symbol of how long you've been out there and how much time away from civilization you have spent immersed in science. It signifies to yourself, your colleagues, and the world that you belong in this role—that you have the ruggedness, the fortitude, and the physical and psychological strength to persist.

The beard explicitly and overtly marks who belongs in the field—both in the sense of engaging in fieldwork and in the broader field of paleontology.

3

These *Bearded Lady* portraits lead us to recognize the obvious: the beard is not a marker that is available or acceptable for women to display. The portraits then ask us to begin to recognize the nonobvious: What other advantages are held by some people and withheld from others?

The beard grown in the field signifies a host of structural advantages, such as a lack of caregiving responsibilities, freedom to travel alone, and economic privilege and security. The question then shifts to the advantages and disadvantages embedded in current-day science: How do institutional and interpersonal practices perpetuate assumptions about who belongs in science? And how can we change these practices to open the doors to science?

The portraits captured in *The Bearded Lady Project* challenge stereotypes about both scientists and the nature of scientific work. A scientist can be male, female, or nonbinary; of any ethnicity; old or young; and the work of science can be done collaboratively; for humanitarian purposes; for the joy of discovery as well as for achievement, for competition, and for status. *The Bearded Lady Project* urges the audience—both scientists and consumers of science—to ask: What are prototypes of scientists in different fields and why? Who is left out of that picture? What is the consequence to science of leaving some people out? Indeed, how is scientific knowledge itself different because of who is constructing it?

Combatting stereotypes requires recognizing them so that we can change the surrounding settings or change individual minds. The pictures in our heads that reflect traditional images may still come forth, unbidden, especially when we are tired or busy. But with awareness we can pause the process, ask further questions, and behave in ways that reflect the values we want to uphold—whether these values include principles of inclusion, of meritocracy, or of prioritizing empirical fact over expectations and norms. With awareness, we can invite a wider range of people into science and eventually change the cultural image of scientists. *The Bearded Lady Project* raises awareness of the historical and ongoing underrepresentation of women in paleontology, and in doing so helps build the foundation of a stronger community for women so that we can form new pictures in our heads.

Amanda Diekman, Ph.D., is a social psychologist who specializes in gender stereotypes, social roles, and how normative beliefs relate to social change and stability. Her research is supported by the National Science Foundation, and she is a Fellow in the Association for Psychological Science, the Society for Experimental Social Psychology, and the Society of Personality and Social Psychology.

Kristine Zellman

Geologist, United States Geological Survey

2

What's in a Name?

Amy K. Guenther

The name "bearded lady" is complicated and often problematic. On the one hand, *The Bearded Lady Project* playfully alludes to the historical performance of the bearded lady figure, most notably in nineteenth-century circuses and sideshows, who disrupted traditional Western ideas about gender.

By temporarily wearing a fake beard, the twenty-first-century women of this project call attention to a history of women disrupting gender stereotypes: women accused of possessing too many "masculine" traits such as an education, a desire for independence, and the pursuit of nonstereotypical gender roles and careers. On the other hand, the actual lived experiences of women with beards today and throughout history reflect very real, often cruel consequences. Moreover, facial hair carries significant meanings for affirming, hiding, and/or denying one's gender identity in a society that (still) strictly enforces expectations with regard to gender. Therefore, we must also examine our use of the bearded lady figure to acknowledge both the powerful symbolism and the oftentimes dangerous lived reality associated with the image at the center of this project.

Though most people associate bearded ladies with nineteenth-century circuses and sideshows, their history actually goes back thousands of years. As early as the fifth century CE, Roman author Macrobius referenced a statue of a bearded Venus on the island of Cyprus in his collection of Roman lore *Saturnalia*. Many early female Christian saints also reportedly possessed beards. According to legend, Saint Galla (sixth century CE) grew a beard after being widowed. The story of Saint Paula (fourth century CE) tells of

A freshly bearded Dr. Carrie Tyler prepares for her portrait.

7

a young virgin who grew a beard to deter a would-be rapist. References to Saint Wilgefortis (fourteenth century CE) relate the tale of a young pagan princess in Portugal who grew a beard to avoid getting married and was subsequently crucified by her father. In England, she was known as Saint Uncumber and was a symbol of women "unencumbering" themselves of husbands. Though the story behind the bearded Venus is unclear, bearded Christian saints tend to be associated with women's refusals to get married and enter the patriarchal order.[1]

Unfortunately, traditional U.S. society today finds the bearded lady no less condemnable for her disruptions of standards of beauty and (white) femininity. For many women, facial hair is a source of shame that they try to obliterate with painful, expensive, and time-consuming procedures. Media and advertising still represent facial hair on women as an imperfection that needs "fixing" since, they suggest, facial hair only grows on men and is a symbol of masculinity. They further assert that a woman's facial hair makes her somehow abnormal, less feminine, and thus less beautiful. Indeed, the presence of facial hair on women disrupts an easily discernible gender binary between men and women. Even the *Oxford English Dictionary* defines beards in conjunction with "an adult man's face." Moreover, the *OED* indicates that (men's) beards symbolize "age, experience, and expertise,"[2] reflecting an explicit and positive association of men, expertise, and beards.

However, there are artists and social media stars attempting to redefine, subvert, normalize, and/or even feminize women with beards. Several bearded performance artists have reasserted their control over the bearded lady narrative in twenty-first-century circuses and sideshows, such as Jennifer Miller, founder of Circus Amok, which specializes in political circus performances. Social media has also given a new visibility to women with hirsutism, a medical condition marked by excessive hair growth, especially on the face and chest, who have chosen to stop removing hair.[3] The most famous perhaps is Harnaam Kaur, who set a Guinness World Record in 2016 for being the youngest woman to grow a full beard and who now works as a model and body positivity activist.[4] Unfortunately, however, social media also has expanded the ways women with beards can be shamed and criticized through hurtful comments, messages, and emails. Increased visibility and positive representations can also bring increased harassment.

Moreover, as LGBTQIA+ people have become more visible, beards and "bearded lady" have taken on even more complicated meanings in terms of gender identity and sexual orientation. For example, historically, "beard" has been used as a slang term to describe a heterosexual relationship used to hide the sexuality of at least one of the partners involved. In this usage, the need for a beard results from living in a homophobic society and/or culture and can have negative emotional repercussions for both parties. Whereas this definition uses "beard" in a metaphorical sense, the physical reality of beards continues to hold very charged connotations for gender expression and gender identity. For instance, some trans women discuss growing large beards before transitioning in an attempt to hide their gender identity from others and, often, themselves. Facial stubble and hair might also cause trans women to experience a psychological condition known as dysphoria, which, in relation to gender, causes "anxiety and/or discomfort regarding one's sex assigned at birth." Facial hair on trans women might also prevent them from being read as women by some. Yet for many trans men, the growth of facial hair can be empowering, gender affirming, and/or part of passing. Facial hair, especially when contrasted with more feminine expressions of make-up and dress,[5] can be used as a way of expressing nonbinary gender identities that are both male and female, neither, or a combination of the two. However, "bearded lady" is still used as a slur for trans people and a derogatory insult for women who can grow facial hair; just as "woman" describes a diverse category of people existing at the intersection of many different identities such as race, gender, and sexuality, there is no one singular way of thinking among people who identify as queer or transgender. For some, "bearded lady" is a term to be taken back, much like the term "queer," while for others it will always be offensive.

The bearded lady, it seems, is still a complicated reality in the twenty-first century. As metaphor, she disrupts society's ideas of what gender is and what a woman should be and challenges patriarchal assumptions. As lived reality, people with facial hair who are not cis men (people whose gender identity [man/male] matches their sex assigned at birth [male]) still face very real social stigmas with material consequences. The name "bearded lady" plays into complicated and problematic histories that have challenged and subverted women but also caused pain and shame and reinforced gender binaries. In demonstrating the ridiculousness of associating expertise, authority, and virility with male facial hair, *The Bearded Lady Project* empowers women in traditionally male-dominated fields and challenges the hyperfeminized standard of beauty for women still portrayed in the media. *The Bearded Lady Project* documentary and its accompanying portraits illuminate the gendered stereotypes and assumptions associated with paleontology, with implications for other academic disciplines. In playing the bearded lady, we look for new possibilities in what facial hair signifies, whom exactly it benefits, and how it disrupts obsolete gender standards.

Amy K. Guenther (she/her) is a freelance scholar, dramaturg, and teacher in Austin, Texas. She has a Ph.D. in theater history, literature, and criticism with an emphasis in performance as public practice from the University of Texas at Austin.

1. For further reading, see Mark Albert Johnston, "Bearded Women in Early Modern England," SEL 47, no. 1 (2007), 1–28; Alexander H. Krappe, "The Bearded Venus," *Folk-Lore* 56, no. 4 (1945), 325–335; and Lewis Wallace, "Bearded Woman, Female Christ: Gendered Transformations in the Legends and Cult of Saint Wilgefortis," *Journal of Feminist Studies in Religion* 30, no. 1 (2014): 43–63.
2. "beard, n.," OED Online: Oxford University Press, accessed September 30, 2018.
3. Roughly 10 perccent of people with ovaries have a medical condition called Polycystic Ovary Syndrome (PCOS). PCOS is the leading cause of hirsutism. Approximately 70 percent of people with PCOS have hirsutism with varying degrees of hair growth. "Polycystic Ovary Syndrome," Office of Women's Health, accessed April 25, 2019, https://www.womenshealth.gov/a-z-topics/polycystic-ovary-syndrome#17; "Polycystic Ovary Syndrome (PCOS)," The American College of Obstetricians and Gynecologists, accessed April 25, 2019, https://www.acog.org/Patients/FAQs/Polycystic-Ovary-Syndrome-PCOS.
4. Meredith Clark, "5 Women with PCOS Explain Why They Choose to Celebrate Their Facial Hair," *Allure*, May 30, 2018, https://www.allure.com/story/women-with-pcosfacial-hair-beard-interviews; Janell M. Hickman, "Instagrammers Challenge Body and Facial Hair Stigma," *Teen Vogue*, March 28, 2017, https://www.teenvogue.com/story/girls-challenging-body-and-facial-hair-stigma.
5. There is no single trans or queer experience. Much of the discourse surrounding these identities happens very fluidly over social media, which would be impossible to adequately cite here. Here are a few of the articles I found helpful in thinking through the complexities surrounding facial hair, gender identity, gender expression, and sexuality: Galen Mitchell, "I Was a Bearded Lady—I Just Didn't Know It Yet," TransSubstantiation, May 16, 2017, https://transsubstantiation.com/i-was-a-beardedlady-i-just-didnt-know-it-yet-1a1ba2b97c59; Dan Avery, "Trans Women Taught Me What a Denial Beard Is," NewNowNext, December 26, 2017, http://www.newnownext.com/denial-beards-transgender-women/12/2017/; "'I Like to Consider Myself Genderful': Interview with Bearded Lady Little Bear Schwarz," Ravishly, accessed July 24, 2019, https://www.ravishly.com/2015/01/07/interview-bearded-lady-little-bear-shcwarz; Natalie Wynn, "Transtrenders," ContraPoints, July 1, 2019, https://www.youtube.com/watch?v=EdvM_pRfuFM.

3

Sex, Science, and Beards

Kimberly A. Hamlin

In 1877, the bearded Dr. Louis Duhring, a founding member of the American Dermatological Association and professor of skin diseases at the University of Pennsylvania Hospital, published an alarming case study about a new disease he and his colleagues feared might reach epidemic proportions.

In "Case of a Bearded Woman," Duhring described a patient named "Viola M." who was unlike any he had ever seen before: a young, healthy mother with a full beard. What confounded Duhring was not so much the thick, dark hair covering Viola's face and neck, but the extent to which she lived an otherwise normal life as a married woman and mother of two. Viola's unusual appearance challenged Duhring's ideas about the "natural" boundary separating women from men and forced him to reconsider what exactly it meant to be female: Could "real" women have beards?

Duhring was not alone in his fascination with the cultural and biological meanings of female facial hair; rather, he was the harbinger of a widespread trend. Between 1877 and 1920, scores of dermatologists reported at conferences and in medical journals that their female patients were traumatized by hypertrichosis, the disease of "superfluous hair." As the doctors debated its etiology and treatment, the public flocked to see bearded ladies on display at circuses and sideshows. From the early 1880s until her death in 1926, the most popular bearded lady was Krao, a woman who had been captured in Laos as a young girl so that she could be exhibited as "Darwin's Missing Link." Unlike Viola, who visited the dermatologist in the hope of removing her beard, Krao became famous for hers. Eventually, she became one of the highest-paid performers employed by impresario P. T. Barnum.

What prompted the public and medical fascination with bearded women? Beginning in the 1870s, women had begun to attend college in record numbers. As the boundaries between male and female spheres started to blur, even only slightly, some doctors argued that this demographic shift had prompted the epidemic of hypertrichosis and perhaps even fueled the public fascination with hairy women. As Dr. J. Herbert Claiborne observed in a medical journal, facial hair on women was caused by "the invasion by woman of many forms of business, professions, trades and heretofore recognized prerogatives of man. I refer in particular to the suffragette feminist movements." Claiborne viewed hypertrichosis as a transitional stage experienced by women who were on their way to becoming men. In his view, bearded men epitomized what it meant to be a scientist; bearded women were considered diseased or looked upon as sideshow freaks.

In fact, women's entry into higher education became the most contentious women's rights issue of the 1870s and 1880s. Could female physiques withstand the rigors of higher education, to say nothing of graduate school or a profession? The nation's leading doctors and scientists said no. Harvard professor and physician Edward Clarke hoped to settle the debate once and for all with his best-selling treatise *Sex in Education or A Fair Chance for the Girls* (1873), which argued that it was physiologically detrimental for women to go to college and menstruate at the same time. Delicate female systems simply could not withstand such pressure, declared the bearded Clarke. Young women who persisted in going to college, he warned, risked becoming infertile. He even predicted that a third gender would evolve: sexless women. Most dangerous of all to female physiology, according to Clarke and his colleagues (including Dr. William Hammond, a founder of the American Neurological Association), was the study of subjects such as science, math, and engineering.

Pioneering female physician Mary Putnam Jacobi successfully refuted Clarke's claims about the taxing nature of menstruation, and female college graduates across the country rallied to publicize their own testimonies of good health. In developing his theory, Clarke, an otolaryngologist, had relied upon secondhand observations of women from his male colleagues. Jacobi, in contrast, conducted the largest study to date of menstruating women, utilizing interviews as well as a variety of new laboratory tests, to determine to what extent menstruation was taxing. Jacobi concluded that menstruating women did not require extra rest and, in fact, that the most active women suffered the least. Her prize-winning essay, "The Question of Rest for Women During Menstruation" (1877), helped define the parameters of the scientific method by establishing that laboratory tests counted more than anecdotes. Nevertheless, elements of Clarke's argument persisted in popular culture, if not in scientific treatises; many people continued to fear that perhaps there was something dangerous about the higher education of women, especially in the sciences.

Since the late nineteenth century, sex, science, and beards have often been discussed in terms of one another. Scientists have argued that men were the only ones capable of scientific research and that in some way this scientific acumen was linked to an inherent quality of manliness, often represented historically by a beard. Women have responded by helping to create the scientific method, by discrediting biological determinism, and by challenging the cultural significance of beards on both men and women.

In the twenty-first century, *The Bearded Lady Project* invites us to confront these gendered contradictions and also think about what it means for women, then and now, to be scientists when the very foundations of science have been constructed in masculine terms.

The Bearded Lady Project reveals the extent to which we associate science with masculinity, consciously or not, and introduces many female paleontologists who have dedicated their lives to helping us better understand the earth and our place on it. By disrupting our assumptions about science and the history of science, *The Bearded Lady Project* also helps us visualize a more egalitarian and a more scientific future.

Kimberly A. Hamlin, Ph.D., teaches history and American studies at Miami University (Ohio). Portions of this essay draw from her award-winning article, "'The Case of a Bearded Woman': Hypertrichosis and the Construction of Gender in the Age of Darwin" (American Quarterly, 2011). She is the author of *From Eve to Evolution: Darwin, Science, and Women's Rights in Gilded Age America* (Chicago: University of Chicago Press, 2014) and the forthcoming book *Free Thinker: Helen Hamilton Gardener's Audacious Pursuit of Equality and the Vote* (New York: Norton, 2020).

Dr. Ellen Currano

Paleobotanist, University of Wyoming

4

What Is Paleontology?

Ellen Currano

"Dinosaurs are big and scary, big animals that aren't very hairy."[1] This is how I remember being introduced to paleontology, as a five-year-old in Mrs. Brizzalera's first grade class at Pope John XXIII School in Evanston, Illinois.

No doubt there were visits to Chicago's Field Museum before that, and my family likes to joke about how I threw a major tantrum at age two during our family vacation to Badlands National Park because my parents told me that I was not allowed to take home what I thought was the very best and most important rock ever. (Looking at our iconic family photograph of me clutching the rock, I now see nothing at all special about that hunk of gray mudstone.) But these lines of a song, learned from a teacher to whom I am forever indebted, are how I remember discovering paleontology. And I became dinosaur obsessed, over the years acquiring dinosaur bed linens, posters, stuffed animals, books, toys, videos, and more, and persuading my obliging parents to take me to as many dinosaur museums as possible. We bought a family membership at the Field Museum, and I dragged my mom on every possible occasion to Dave's Rock Shop in our neighborhood, which also has quite an impressive museum in its basement.

While I knew that other kinds of fossils besides dinosaurs existed (and I saved my allowance and babysitting money to buy ones I fancied from Dave), I don't think I really understood that there were paleontologists who studied anything other than dinosaurs until I went to college. True, I had several fossil books with beautiful photographs of petrified wood, ammonites, crinoids, and more, and my mom read me excerpts from the University of Chicago alumni magazine about new discoveries by world-renowned invertebrate paleontologists who would become my professors when I too attended university there. But I thought that people studied these fossils in their free time—when they weren't studying dinosaurs. Oh my, how wrong I was, and how much more wonderful my world is for discovering it!

Paleontologists study the history of life on Earth, from the very first single-celled organisms to our own hominid ancestors. We are often confused with archeologists, who study human history and civilization. Paleontologists seek to understand how ancient and extinct organisms functioned and what environments different organisms inhabited. Some of us could just as well be called evolutionary biologists because we use fossils to study how evolution occurs and how different groups are related to each other. We discover the "missing links," or, more correctly, transition series, that show intermediaries between seemingly disparate groups: feathered dinosaurs, legged whales, and, near and dear to me at least, trees with wood like a conifer but leaves and spores like a fern. Others of us focus on the ecology of ancient Earth: what organisms are found together, how they interacted with each other, and how ecosystems responded to environmental perturbations. Thus, in addition to being a gateway science that, through dinosaurs, lures in unsuspecting children, paleontology is a societally relevant science.

Today humans are causing the extinction of plants and animals and global warming, the likes of which are unknown in recorded history. By studying fossils, we can establish a baseline for how Earth worked before there were people. We can investigate the rates and magnitudes of past climate change and mass extinction, the resilience of different types of organisms, and how life previously rebounded from catastrophe. It turns out that what humans are doing is unprecedented in the 4.6 billion-year history of Earth.

The scientists of *The Bearded Lady Project* encompass the breadth of paleontology. I study ancient plant life, and my research investigates how plants and the insects that ate them responded to ancient climate change. I do this by excavating fossil leaves and examining them for traces of insect feeding damage. Similarly, invertebrate paleontologist Patricia Kelley studies predator-prey coevolution and escalation over the last 80 million years by examining holes in bivalve shells made by boring snails and other predators. Paleoecologist Karen Chin studies food webs, which include dinosaurs, by mounting very thin slices of fossil dinosaur dung on microscope slides and identifying everything she can within the dung. Other paleontologists use fossils to document ancient environmental disturbances. Andrea Hawkes and Tina Dura collect sediment cores from marshes in places that are prone to earthquakes and tsunamis, isolate diatoms and foraminifera (two types of microfossils) from the sediment, and use species composition and

Fossil fruit (left) and leaf (right) of an extinct species of sycamore that was discovered in circa 50 million-year-old rocks in the Wind River Basin, Wyoming

abundances to reconstruct sea-level changes caused by tectonics. Their goal is to extend our record of earthquakes and tsunamis back thousands of years and use these data to better assess hazards. Claire Belcher's lab focuses on the history of wildfire, examining charcoal that is preserved in ancient sediments. She and her colleagues also conduct experiments in the laboratory, burning different plant material to determine flammability and carefully observe features of the charcoal produced. And for all paleontologists, it is essential to understand how well fossil remains capture the living communities. This is the life's work of paleobiologist Anna K. ("Kay") Behrensmeyer, cofounder of an entire discipline within paleontology called taphonomy, or the study of how remains become fossilized. Behrensmeyer has been studying carcasses at Amboseli National Park in Kenya for decades to better understand the processes that preserve and destroy modern bones, as well as to investigate how faithfully the bone assemblages capture what is actually living in Amboseli.

Paleontology has come a long way since the early 1800s, when Mary Anning excavated the fossil bones of then unimaginable animals from the cliffs of Lyme Regis in England and had to turn them over to men for scientific study. Today, women can study and publish on the fossils that they find, but, as in other fields in science, technology, engineering, and mathematics (STEM for short), women remain severely underrepresented.

Just 22 percent of Paleontological Society professional members are women, a proportion that has not changed since 2000.[2] Broadening to the geosciences, women make up 20 percent of faculty members.[3] Recognition of female paleontologists also lags. The median annual salary for a man with a Ph.D. in geoscience is $98,000, versus only $80,000 for a woman.[4] The top award given by paleontology's largest professional society, the Paleontological Society Medal, has been awarded to just four women, versus 55 men. But encouragingly, the Paleontological Society is increasingly recognizing women for early career excellence; between 1973 and 2012, just three women received the society's Schuchert Award, whereas four of seven awardees since 2012 are women.

A steadily growing number of studies seek to understand why so few girls become scientists and why the retention rate for female scientists is so low. There is no evidence for innate differences in scientific aptitude between men and women.[5] Rather, the most recent work argues that implicit and likely unintentional biases that stem from repeated exposure to cultural stereotypes prevent many talented women from succeeding.[6] Bluntly put, women are perceived as less competent but more "likeable" or "warm" than men.[7] A now-famous study by Corrine Moss-Racusin and colleagues conclusively demonstrated that both women and men display biases against aspiring female scientists. Female applicants for lab manager positions were less likely to be hired, offered a lower salary, and viewed as less deserving of mentorship than otherwise identical male applicants.[8] Similarly, a study of reference letters for geoscience postdoctoral fellowships, a key stepping-stone to coveted faculty positions, found that women are half as likely as men to receive excellent letters.[9]

These biases affect women throughout their careers. In addition to being overlooked for jobs, awards, leadership positions in professional societies, and other professional honors, women in STEM are vulnerable to stereotype threat:[10] the constant pressure of having to be perfect, so as not to reinforce negative stereotypes about women in science. One manifestation of stereotype threat is lowered self-confidence, which in turn leads to greater attention to detail, fewer grant applications and less ambitious requests, and lower publication rates.[11] Grant dollars and publication record are key metrics used to evaluate prospective faculty hires and during the promotion process.

The Bearded Lady Project (1) challenges societal messages and the implicit stereotype of lower scientific competence in women; (2) showcases, as role models, some inspirational, field-based paleontologists who also happen to be women; and (3) provides opportunities for aspiring scientists to conduct their own field-based research through our scholarship fund (see page 196).

The Bearded Lady Project began before #MeToo, but our public events have initiated discussions on sexual harassment and assault, both of which are prevalent in science and especially in field-based sciences. Like our sisters in entertainment, politics, sports, and other male-dominated industries, each and every woman in paleontology has #MeToo stories to tell. It has been an honor to provide a venue in which women can be empowered by publicly sharing experiences that they have kept hidden for too long. *The Bearded Lady Project* team proudly captures both the struggles and the triumphs of women in paleontology.

Ellen Currano, Ph.D., is an associate professor of paleobotany at the University of Wyoming and a cofounder of *The Bearded Lady Project*.

1. Google tells me that the line is actually, "Oh, Dinosaurs were strong and scary! / Big animals who are not very hairy!!!" and that it is from the rock opera *Dinosaur Rock*.
2. Phoebe A. Cohen, Alycia Stigall, and Chad Topaz, "A Gender Analysis of the Paleontological Society: Trends, Gaps, and a Way Forward," poster at North American Paleontological Convention, 2019.
3. Carolyn Wilson, *Status of the Geoscience Workforce* (Alexandria, VA: American Geosciences Institute, 2018).
4. National Science Foundation and National Centre for Science and Engineering Statistics, *Scientists and Engineers Statistical Data System Surveys: Survey Year 2013*.
5. Elizabeth S. Spelke, "Sex Differences in Intrinsic Aptitude for Mathematics and Science? A Critical review," *American Psychologist* 60, no. 9 (2005): 950–958; Janet S. Hyde and Marcia C. Linn, "Diversity—Gender Similarities in Mathematics and Science," *Science* 314, no. 5799 (2006): 599–600; Diane F. Halpern, Camilla P. Benbow, David C. Geary, Ruben C. Gur, Janet S. Hyde, and Morton Ann Gernsbacher, "The Science of Sex Differences in Science and Mathematics," *Psychological Science in the Public Interest* 8, no. 1 (2007): 1–51.
6. Corrine A. Moss-Racusin, et al., "Science Faculty's Subtle Gender Biases Favor Male Students," *Proceedings of the National Academy of Sciences of the United States of America* 109, no. 41 (2012): 16474–16479; Calvin K. Lai, Kelly M. Hoffman, and Brian A. Nosek, "Reducing Implicit Prejudice," *Social and Personality Psychology Compass* 7 (2013): 315–330.
7. Alice H. Eagly and Antonio Mladinic, "Are People Prejudiced Against Women? Some Answers from Research on Attitudes, Gender Stereotypes, and Judgements of Competence," *European Review of Social Psychology* 5, no. 1 (1994): 1–35; Moss-Racusin, et al., "Science Faculty's Subtle Gender Biases Favor Male Students," 109.
8. Moss-Racusin, et al., "Science Faculty's Subtle Gender Biases Favor Male Students," 109.
9. Kuheli Dutt, et al., "Gender Differences in Recommendation Letters for Postdoctoral Fellowships in Geoscience," *Nature Geoscience* 9 (2016): 805–808.
10. Claude M. Steele and Joshua Aronson, "Stereotype Threat and the Intellectual Test Performance of African-Americans," *Journal of Personality and Social Psychology* 69, no. 5 (1995): 797–811.
11. Matthew R. E. Symonds, et al., "Gender Differences in Publication Output: Towards an Unbiased Metric of Research Performance," *Plos One* 1, no. 1 (2006): e127.

5

Spaces Paleontologists Inhabit

Ellen Currano

Many of our Bearded Ladies became professional paleontologists because they did not want to spend every workday inside, in an office, behind a desk.

Some were excited to have a job that varies throughout the year, from days spent working outdoors and nights camped under the stars to days using laboratory equipment and computers. Popular media portrayals of paleontology tend to focus on fieldwork, as this is viewed as the most beautiful and exciting part of the job. Yet the field is just one space that paleontologists inhabit. The text and photographs below describe the four places you are most likely to find a paleontologist, providing a more holistic picture of her professional life.

1. The Field

"You're a paleontologist, so you must go on expeditions and dig up fossils, right?" Paleontologists featured in *The Bearded Lady Project* have conducted fieldwork on all seven continents, from remote badlands on the other side of the globe to roadcuts that are practically in their own backyards. Some even forgo working on the continents and instead conduct field research aboard ships in the middle of the ocean. A paleontologist chooses her field sites based on the scientific questions that drive her research. Where are there sedimentary rocks of the right age exposed? Do these rocks preserve the right ancient environment? Have any fossils previously been discovered there? Paleontologists have also been known to think of interesting research questions in order to visit a dream location. And of course, personal safety and ability to obtain the necessary permits and permissions must also be considered.

Drs. Claire Belcher and Sarah Baker search for fossils in the cliffs of Mary Anning's field site in Lyme Regis, West Dorset, England.

What a paleontologist does in the field varies considerably. *Vertebrate paleontologists* scour the ground for fossil remains, collect giant bags of sediment and screen wash them to isolate tiny bones and teeth, or spend weeks excavating giant dinosaurs or mammals. *Micropaleontologists* core: mud flats, lakes, ponds, and oceans. *Paleobotanists* wander the badlands, by truck and by foot, looking for rocks of the right color (drab brown or gray), and then dig small test quarries to see whether plant remains are preserved in the rock. If so, large quarries are opened, from which a hundred leaves, fruits, and seeds can be recovered in a single day and wrapped in toilet paper for safe transport. Similarly, once an *invertebrate paleontologist* finds a productive fossil locality, she can discover tens to hundreds of fossil creatures in a day. She might collect only those fossils that are beautifully preserved or buckets of rock or sediment in order to identify and count each and every fossil back in the lab. It all depends on the research question.

2. The Lab

Data collection begins in the field but continues in the laboratory for many months and even years after fieldwork ends. *Vertebrate paleontologists* return from the field with their fossils wrapped in plaster casts and still partly encased in rock. *Micropaleontologists* have sediment cores or golf ball–sized samples of rock or mud, from which pollen grains and minuscule marine organisms (most often foraminifera and plankton) are isolated and mounted on microscope slides for observation. *Paleoecologists* have hundreds and even thousands of fossils that must be sufficiently cleaned to be identified and counted. To understand how organisms work and how different groups are related to each other, paleontologists measure aspects of the size and shape of fossils. To study the ecology of ancient Earth, paleontologists scrutinize every scrap of fossil material from a site, both to tabulate the number of individuals of each species and to search for evidence of interactions among organisms (e.g., insect feeding traces on fossil leaves). And once these data are collected, paleoecologists use statistical analyses to investigate differences among samples. Then it's a matter of searching the scientific literature to understand how the new results complement what has already been done, writing the research up for publication in scientific journals, and writing grant proposals to fund the next field season.

Most paleontology labs include benchtop microscopes, cameras, computers, and specimen cabinets. However, paleontologists often share lab equipment with other scientists or bring fossils to external labs for research. Mass spectrometers are used to study the chemistry of fossil remains, scanning electron microscopes provide images of the tiniest details of the

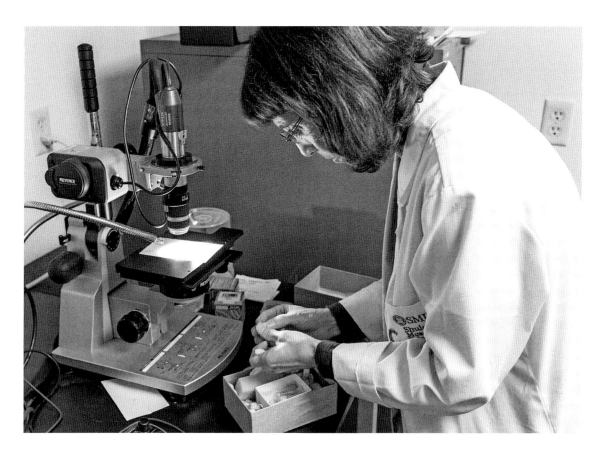

Dr. Alisa Winkler in her lab at Southern Methodist University, Dallas, Texas

surface of fossils, and CT scanners capture the 3D structure of fossils without having to isolate specimens from their encasing rock matrix, particularly useful for complex and hollow remains like skulls. Some paleontologists even visit a cyclic particle accelerator, or synchrotron, to obtain the very highest resolution information about their fossils.

3. The Classroom

Colleges and universities are the main employers of paleontologists, as professors in geology, biology, or environmental sciences departments. In addition to research, job duties include teaching, mentoring undergraduate and graduate student researchers, and service. The courses a paleontologist teaches vary depending on her background and the type of institution at which she works, but usually include introductory or general education courses in biology or geology, upper-division undergraduate paleontology courses, and specialized graduate seminar courses. *Vertebrate paleontologists* can even be employed by medical schools to teach anatomy. For every hour in the classroom directly interacting with students, there are

23

many hours spent behind the scenes designing the course; writing lectures, projects, and exams; grading; holding office hours and review sessions; responding to student emails; and writing letters of recommendation.

Teaching college-level science classes poses particular challenges for women. Student evaluations, which at many institutions are an important component in evaluating faculty for promotion and tenure, are well known to be biased against women. This was brilliantly demonstrated in a study conducted using an online course.[1] Four sections of the same course were offered, two taught by a female instructor and two by a male instructor. The female instructor used her own name in one section and the male instructor's name in the other; the male instructor did the same. Students were asked at the end of the semester to complete evaluations, and, regardless of who actually taught the course, the perceived male instructor received a significantly higher overall instructor rating, as well as higher scores for professionalism, fairness, promptness, respectfulness, enthusiasm, and giving praise. Here, as in other aspects of a paleontologist's job, women must work harder than men to achieve the same success. And they must

Dr. Kate Bulinski leads a discussion during a field trip at the Falls of the Ohio State Park, Indiana.

24

put time and energy into what to wear to class because, unlike for male professors, discussions of personal appearance and clothing choices routinely appear on student evaluations.

4. The Museum

Many people's first time seeing a fossil is during a childhood visit to the local natural history museum. An extremely important work activity for a paleontologist employed by a museum is to collaborate with science educators, artists, and exhibit designers to create exhibits that use fossils to teach children and adults fundamental scientific concepts, including evolution, climate change, plate tectonics, extinction, and Earth history. Paleontologists commonly participate in public education programming and are on call to identify potential fossils brought in by visitors.

Paleontologists also play an extremely important role behind the scenes at museums, as curators of the fossil collections. Fossils are the primary data on which paleontological research is built—each specimen records the occurrence of a particular organism at a given place at a given time. This information must remain associated with each fossil, and both pieces made available to scientists and students for research. Of particular importance is a museum's type collection. Whenever a new species is named, one specimen that best shows the distinguishing characteristics of that species is selected to be the holotype. Holotypes must be kept in a public repository for others to examine. Paleontologists are continually discovering new methods to study fossils and asking new questions about Earth's past. Fossils discovered hundreds of years ago are still useful for research, and it is the responsibility of curators to ensure that fossil collections are preserved properly for tomorrow's scientists.

Ellen Currano, Ph.D., is an associate professor of paleobotany at the University of Wyoming and a cofounder of *The Bearded Lady Project*.

---

1. Lillian MacNell, Adam Driscoll, and Andrea N. Hunt, "What's in a Name: Exposing Gender Bias in Student Ratings of Teaching," *Innovative Higher Education* 40, no. 4 (2015): 291–303.

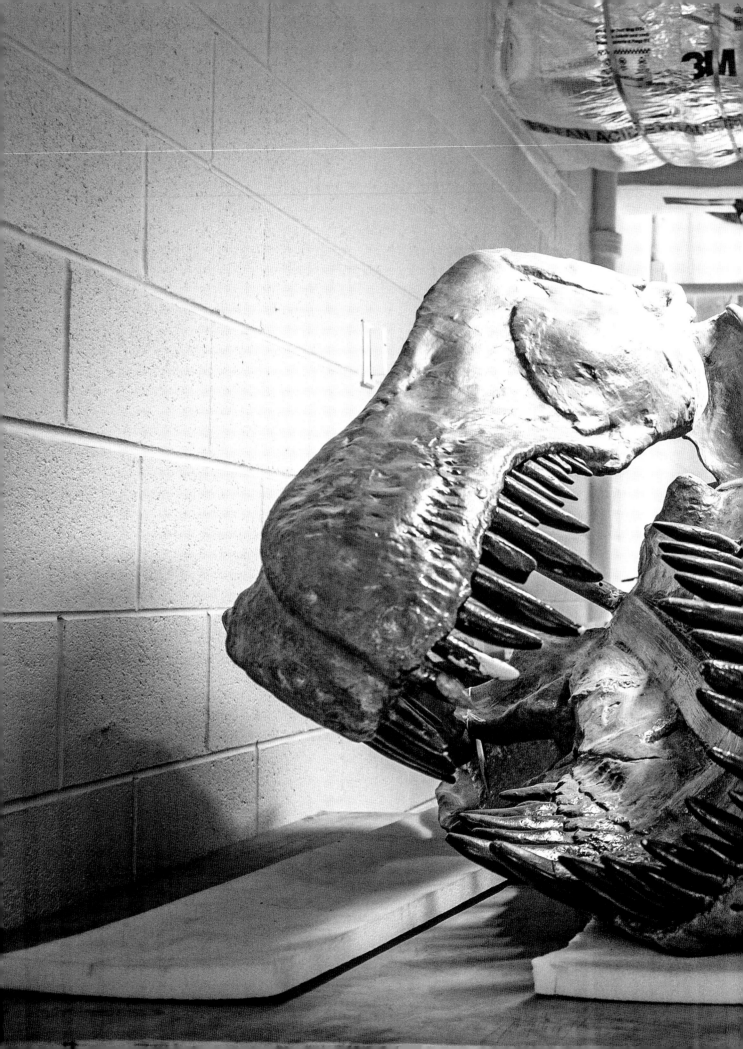

6

The Lost Legacy

Ellen Currano

Black-and-white portraits of past paleontologists continue to make a lasting and often outdated impression on the field of paleontology. These images are memorable and stereotypical: male colonizers standing beside their dinosaur bone trophies, their unkempt beards badges of honor for time spent in the field.

We know this image well because the contribution of these white male scientists lingers with their photographs staking claim on their discoveries. The bearded faces are proof that they were there.

But what about the contributions made by women? Women who were in the field, who discovered crucial fossils, and whose work pushed paleontology forward—where are their photographs? The lack of images documenting women in paleontology erases them from the historical landscape, and from our classroom discussions and textbooks. *The Bearded Lady Project* inserts women into the historical narrative by physically putting female scientists in the bearded portraits that are the norm upon which we base our expectations of what a paleontologist should look like. The truth is, women in paleontology exist, but their legacy was hard fought and remains in the shadows.

Here we introduce four women who made important contributions during the early days of paleontology but are not mentioned in textbooks. Their framed portraits do not hang on the walls of science classrooms and university buildings. These scientists, who represent a small fraction of the lost legacy of women in paleontology, were chosen to highlight the varying roles women have played in paleontological research.[1]

Mary Anning

From an early age, Mary Anning (1799–1847) made a living scouring the cliffs of Lyme Regis, England, for Jurassic fossils and selling them.[2] It was dangerous work; she had to correctly time the tides, navigate steep and slippery shores, and avoid mudflows and landslides. Her beloved dog Tray, constant companion on her fossil hunts (the original paleontological field dog?!), was killed when a cliff collapsed and buried him. She discovered and carefully excavated ichthyosaurs, plesiosaurs, and even a pterosaur, and these discoveries revolutionized paleontology. Anning's observations and her self-taught knowledge of anatomy convinced her that these fossils were species no one had ever seen alive. Today extinction is universally recognized, and nearly all paleontologists do research that can be somehow tied to extinction. When Anning lived, however, most people believed in the biblical record of Earth's creation and had no means to explain how there could be species from Earth's past that were not still present. Extinction implies imperfect creation, which was difficult to reconcile with early 1800s religious doctrine. Anning was not allowed to publish on her finds because she was a woman, and even worse, a working-class woman. Many of the great male geologists and anatomists of her day published her work about specimens, though few acknowledged her as the collector. It is only in recent years that Mary Anning has been celebrated as the greatest fossil hunter ever known (the Natural History Museum of London) and one of the ten most influential British female scientists (the Royal Society).

Winifred Goldring

Winifred Goldring (1888–1971) shattered two glass ceilings by being the first woman appointed state paleontologist and the first woman elected president of the Paleontological Society. She spent her entire career at the State Museum of New York, beginning in 1914 as a scientific expert in paleontology and working her way up to state paleontologist by 1939. When she retired in 1954, she had published forty-four papers that spanned a breadth of geological and paleontological topics and established herself as an expert on the geology of New York State, crinoids, and early land plants. In addition to her academic pursuits, Goldring was passionate about the importance of museums for public education. She replaced exhibits that were simply glass boxes of curiosities with exhibits that used fossils to tell a story about some aspect of natural history. She also created some of the first dioramas that reconstructed entire ancient habitats. While Goldring herself is rarely mentioned in textbooks, photographs of her diorama of Gilboa, a 385 million-year-old fossil site in New York, which preserves the earliest forest, appear prominently.[3]

Yusra

In 1932, Yusra discovered Tabun 1, a nearly complete skull of a female Neanderthal, at the Mount Carmel site in Israel. Had this find been made by a white, formally trained scientist, it would have led to fame and a long scholarly career. But Yusra, whose last name is unknown, was one of the many local women of color who were recruited to help with fieldwork. Excavations at Mount Carmel were led by British archeologist Dorothy Garrod, a trailblazer in her own field (research focusing on the fossil bones of human ancestors can equally be considered archeology and paleontology). The site became a hub for archeologists, paleontologists, and geologists who were women, many of whom went on to successful scientific careers. Others remain unknown. Women like Yusra who, over the six years of the project, became adept field collectors and made sensational discoveries, rarely had such opportunities. Yusra dreamed of one day studying at Cambridge. She never did. What did become of Yusra remains unknown.[4]

Annie Montague Alexander

The University of California Museum of Paleontology (UCMP) has played a prominent role in *The Bearded Lady Project*, so it is fitting to include the museum's founder, philanthropist and field scientist Annie Montague Alexander (1867–1950). Alexander discovered paleontology in 1901 when she attended a lecture at Berkeley. Her father had made a fortune in the Hawaiian sugar industry, and Alexander put that money to good use, bankrolling numerous scientific expeditions, making monthly contributions to support paleontological research at Berkeley, and, in 1934, establishing UCMP. Alexander was outstanding at field collecting and contributed over 20,000 specimens of fossils, plants, and animals to the UC museums. Seventeen species are named after her. She played an active role in all aspects of fieldwork, from funding expeditions to collecting specimens to doing most of the cooking. She appreciated the fact that proper curation of specimens would allow for future research, and the museum she founded remains an important resource for paleontologists worldwide. A memorial stone on the University of Colorado campus reads: "Annie Montague Alexander. She found men a nuisance on her arduous field trips."[5]

What legacy of today's paleontologists will remain one hundred years from now? Will future generations see and be inspired by color photographs that depict diverse twenty-first-century women, in their element, doing scientific research? Will these images make it into textbooks and onto the walls of science classrooms? Will women also gain equal footing in the realm of fiction, which has the power to create role models like Dana Scully who inspire youths to pursue and excel in scientific careers?[6]

It is up to all of us to create a legacy of diversity and inclusivity in science. Teach the work of women and people of color, but also make a point of saying their names and showing their photographs. Feature diverse scientists in the popular media and create Wikipedia pages for them. Create fictional scientists who do not conform to the dorky white male stereotype. Most important, question the images you see, be aware of the lack of diversity, and speak up about it.

Ellen Currano, Ph.D., is an associate professor of paleobotany at the University of Wyoming and a cofounder of *The Bearded Lady Project*.

1. For a more extensive list and photographs of these scientists, we highly recommend the Trowelblazers website (https://trowelblazers.com) by Brenna Hasset, Victoria Herridge, Suzanne Pilaar Birch, and Rebecca Wragg Sykes, and the book by author Robbie R. Gries: *Anomalies-Pioneering Women in Petroleum Geology: 1917–2017* (Lakewood, CO: Jewel Publishing LLC, 2017).
2. Anning might be the inspiration for the tongue twister "She sells seashells by the seashore."
3. For a more extensive biography of Dr. Goldring, see Donald W. Fisher, "Memorial to Winifred Goldring 1888–1971," *Memorials of the Geological Society of America* 3 (1974): 96–107.
4. Information on women of color who were active in the earliest days of paleontology is elusive. Did they not exist, or were they even more likely to be forgotten by history than their white counterparts? After much searching, we discovered the story of Yusra on the Trowelblazers website: https://trowelblazers.com/yusra-expert-excavator-of-mount-carmel/.
5. A good starting place to learn more about Annie Montague Alexander's remarkable life is the biography "Annie Montague Alexander: Benefactress of UCMP," University of California Museum of Paleontology, accessed October 16, 2019, https://ucmp.berkeley.edu/history/alexander.html.
6. Dana Scully, one of the lead characters on *The X-Files*, a popular science-fiction show during the 1990s, was a University of Maryland physics major, graduated from Stanford medical school, and then was recruited by the FBI. She was widely recognized for using the scientific approach to examine the paranormal. According to research by 21st Century Fox, the Geena Davis Institute on Gender in Media, and J. Walter Thompson Intelligence, Scully inspired a generation of women to pursue careers in science: https://seejane.org/research-informs-empowers/the-scully-effect-i-want-to-believe-in-stem/.

7

The Power of Contradiction

Catherine Badgley

The bearded lady: a puzzle of gender identity, amusing and unnerving, authoritative omen of a topsy-turvy social order, eerie and fascinating.

These reactions in audiences that beheld a bearded lady at a nineteenth-century circus are no less relevant for the audiences who view the portraits in this book. The images here invite us to question, even as we smile at the transformation of our friends and colleagues.

I encountered the idea for this collaborative project between paleontologist Ellen Currano and filmmaker Lexi Marsh while reviewing a grant proposal by Ellen. Her proposal focused on new goals for research about early forests of flowering plants, particularly the responses of trees to leaf-eating insects during episodes of climate change from millions of years ago. In addition to fieldwork and analysis, the proposal contained goals for science communication and public outreach, including the formative stage of *The Bearded Lady Project*.

My immediate reaction was: "What a brilliant idea!" (The research goals of the proposal were impressive as well.) The portraits, accompanied by interviews with the scientists, would recast the contradictions of the bearded ladies of past centuries in the context of women changing the norms of academic science. The project would showcase fascinating aspects of paleontology along with challenges that women face entering a workforce that has been dominated by men since its inception. The choice to use an older film technology—a large-format film camera on a tripod—by photographer Kelsey Vance would link the portraits with the camera technology of the nineteenth and early twentieth centuries—the same era when bearded ladies were on display in circus sideshows. The portraits could amuse,

Dr. Catherine Badgley shares her find with Cinematographer Draper White.

Dr. Catherine Badgley
with a favorite field tool:
an ice pick

inspire, start conversations, and challenge old assumptions—what better goals for art to achieve?

I reviewed the grant proposal in 2014, just as the filming of Ellen had begun and planning for additional people and field sites was under way. When I saw Ellen at a scientific meeting later that year, I congratulated her on *The Bearded Lady Project*. My own research in the Mojave Desert focused on Miocene sequences that preserved mammal faunas during the last major interval of global warming, 17 to 14 million years ago. Two doctoral students, Tara Smiley and Katharine Loughney, were working with me. Our goals were to determine how the warming interval was recorded in the ancient Mojave environments and how mammals responded to these environmental changes. Through Ellen, I invited the film crew to visit our field area during our next field season to capture the harsh beauty of the desert landscape and to film female vertebrate paleontologists at different stages in their careers. My students and I were pleased to contribute to an innovative project that would start many conversations in the future. The film crew visited us in Rainbow Basin, near Barstow, California, in April 2015.

While putting on the beard for my portrait, I felt amused, surprised, and puzzled. I was the same person as minutes before, only now wearing a theater prop. On the other hand, this change in appearance transformed us into symbols of women in science. My reaction was stronger when I saw the portraits on exhibit at the Geological Society of America meeting in Seattle in October 2017. The bearded ladies had such bold poses; they

Vertebrate fossils
discovered in the Mojave
Desert

were both familiar and eerily different. Some colleagues who have known me for decades said that they did not recognize me in my bearded portrait. The diverse reactions from viewers illustrated the power of this project.

Our experiences becoming bearded ladies in the twenty-first century have caused us to reflect on the challenges that women face in securing a job and establishing a scientific identity. Although we have more opportunities and better support than even a generation ago, many legacies of the past remain. It is still important to raise and discuss the contradictions that the bearded ladies present to us. The rich portraits of all the bearded ladies link the legacies of the nineteenth century—social and technical—to the contradictions and changing faces of women in science today. A brilliant idea!

Catherine Badgley, Ph.D., is a paleoecologist in the Department of Ecology and Evolutionary Biology at the University of Michigan.

PART II
Women in Paleontology

Stereotypes

Combatting stereotypes requires recognizing them so that we can change the surrounding settings or change individual minds. The pictures in our heads that reflect traditional images may still come forth, unbidden, especially when we are tired or busy. But with awareness we can pause the process, ask further questions, and behave in ways that reflect the values we want to uphold.

Amanda Diekman

Dr. Carole S. Hickman

Invertebrate paleontologist, University of California, Berkeley

8

"Fitting In": Freedom in the Field

Carole S. Hickman

Ten years into retirement, I find that field exploration continues to feed a lifetime of curiosity and discovery. My love affair with the natural world began as a child—collecting butterflies, plants, and fossils as part of observing rich detail, encountering the unexpected, and seeking explanations. Curiosity became increasingly seductive as my education in science revealed so many tools for pursuing explanations. Paleontology offered the best possible opportunity to seek both geological and biological knowledge. Three degrees led to my dream appointment in 1977 to the UC Berkeley faculty in the only Department of Paleontology in the country. Women were anomalies in science at that time, and I was the first woman faculty appointment in the building, which also housed the departments of geology and geography. I took great satisfaction in exceeding university expectations, being awarded tenure early, and setting off for a sabbatical year in Australia in a scientific culture wherein women were also anomalies.

It was during that sabbatical in Australia when the story of how I came to wear a mustache originated. I rented a car to head off into the field on my own in Queensland, and several male colleagues in Sydney warned me that I would be traveling on some dangerous roads and encountering some tough characters. And so I purchased a fake Aussie bloke mustache—to fit in with the excessively hairy upper lip Aussie stereotype. Afterward, I wore it at Halloween, disguising myself as the archetypical macho field geologist. It was fun to repurpose it for *The Bearded Lady Project*, achieving a new dimension of the idea of *challenging the face of science*.

The original idea of fitting in does have a practical rationale. Even during the Victorian era, there were a few bold women who ventured into the field and made important discoveries. However, conventional dress constrained them to corsets and long skirts that were uncomfortable, impractical, and likely to attract unwanted attention. In Australia I learned of anomalous

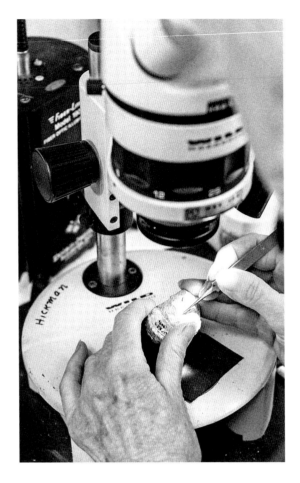

Examining fossil mollusks in Hickman's lab

women in the early twentieth century who did geological and paleontological fieldwork dressed in practical clothing designed specifically for tough field conditions. After all, Victorian men had not gone into the field wearing high-waisted pants, top hats, frock coats, and button-up shoes. Fitting in should always be a matter of comfort and practicality.

In my case, fitting in was also a matter of being inconspicuous and avoiding attention. It was useful not only when working alone but also on otherwise all-male scientific expeditions and postconference field excursions where we shared the same experiences, got grubby, and needed lots of pockets and sometimes very large backpacks for specimens.

The field offers an exhilarating freedom to go where no one else has gone before and to discover what no one else has seen. It is where the excitement begins—the moment when the brain lights up with new connections. Although these "aha moments" also happen in the lab, they are especially exciting when they happen off the main road, on paths less traveled. In my own career there were certainly roadblocks, but deviating from the main road was always the very best part of the quest. The Aussie mustache is one of many mementos of fieldwork, along with a practical Australian bush hat, field notes, journals, photographs, drawings, and specimens of more than 100 new species of living and fossil mollusks that I discovered, named, described, and published research on.

Carole S. Hickman, Ph.D., is a professor of the Graduate School in Integrative Biology and Curator of Invertebrates at the University of California Museum of Paleontology.

Dr. Anna K. ("Kay") Behrensmeyer

Vertebrate paleontologist, Smithsonian Institution

9

A Female Paleontologist in the 1970s and 1980s

Anna K. ("Kay") Behrensmeyer

When I arrived at UC Berkeley in 1973 for a two-year postdoctoral fellowship, one of the surprises was that I had to be careful about what I said around the professors and students. I'd always been conscious of the need to temper "girlish enthusiasm" in order to be taken seriously as a female graduate student at Harvard. That meant I was more of a listener than a talker, partly because I had always been naturally shy. It wasn't all shyness—I listened more than I talked because of the fear of being judged more critically than my male peers. With my doctoral degree in hand, I thought maybe I could express my thoughts and ideas more spontaneously, especially in laid-back California.

At first, people expected me to know more on diverse paleontological topics than I actually did and to talk about these subjects with some authority. They also expected me to speak up and join in, with no sense of gender bias, and they were genuinely interested in my knowledge and opinions. So I had to shift gears into a new, more assertive and confident mode in the Berkeley academic scene. It was a different balance of standards and expectations than I had been used to in graduate school.

Fortunately, it didn't take long for the "settling in" phase to pass, and I think the expectations and discomforts I sensed had more to do with my shiny new East Coast doctorate than with being a female paleontologist in a male-dominated profession. Once my new colleagues learned about my work and I learned about theirs, we found common ground and dropped the pretenses. From then on, life at Berkeley was great. There were a lot of women grad students in both paleontology and anthropology, and for the first time I had female as well as male peers. Some of us bonded over common interests in taphonomy—how animal and plant remains are preserved and what that means for interpreting the fossil record. We had a lot of fun together, composing songs about petrifaction and dreaming about alternative careers advising people on how to become fossilized. I don't recall any barriers or biases that affected me during those years, and I didn't realize until much later how lucky I was to land in one of the early centers of gender equality, with senior (mostly male) colleagues who were committed to fairness.

I had an office in the UC Museum of Paleontology but was affiliated with the African archaeological group led by Glynn Isaac across campus. It was exhilarating to walk through the eucalyptus groves on bright Californian mornings between two exciting communities of paleontologists and archeologists. My dream of being an interdisciplinary scientist—part geologist, part paleontologist—seemed possible, though there was still the problem of landing a permanent job. I had a wonderful sense of freedom and opportunity; it was a great time to unleash my curiosity about taphonomy and how it had shaped the fossil record.

As a postdoc, I got involved in many projects that have carried on throughout my career. Berkeley colleagues and I started transport experiments with bones in flumes and natural river systems and taphonomic surveys of carcasses in local ranches and reserves. I co-led a project on the sedimentary context of Morrison Formation dinosaurs in the western United States, explored modern sedimentary environments in East Turkana, Kenya, and—best of all—found an ecosystem in East Africa where I could study the early stages of bone transitions from biosphere to lithosphere. A temporary lectureship at UC Santa Cruz gave me teaching experience, while I continued research in Kenya. To save money, I lived in a repurposed chicken house on a farm outside of Nairobi, making new friends and reveling in a feeling of self-sufficiency. There were moments when I wondered whether I would ever land a permanent job or find a life partner who could cope with being attached to a female scientist, but I was too busy with new ideas and field projects to dwell much on the uncertain long-term future.

When the time at Santa Cruz ended, I moved to Yale to join a research team working on Miocene fossils and fluvial rocks of Pakistan. All in all, I spent seven happy years as a postdoc. I applied for two jobs during that time, one of which I got. For the other, at the University of Wyoming, I was competing with a more traditional male paleontologist. There may have been some "old guard" bias during my interview, but I also had friendly supporters and positive feedback. After I was turned down, it didn't occur to me that my gender had been an issue; I think they just decided on another kind of paleontologist. For the Smithsonian's National Museum of Natural History position, I was one of two female finalists—a point the chairman made to me in order to quell any notion that I might have been chosen because I was female.

Arriving at the National Museum of Natural History in D.C. in 1981 was a dream come true. I was warmly welcomed, although I do remember one not-so-great experience: the then director of the museum took me out for lunch, which was very nice, but it happened to be on National Secretary's Day. He was very embarrassed when one of his colleagues passed by with a comment assuming that I was his secretary. Such happenings were rare, fortunately. Instead, I felt empowered by my new position and exhilarated to be in an environment where natural history research, collections, and education were all part of the curatorial mission.

It was not so easy to navigate the social life of the D.C. area as a female professional scientist, but the museum provided regular Friday beer gatherings, and there were many new curators—a cohort ready to have fun, advance science, and shake up museum traditions. There were opportunities for leadership in research and educational outreach. It was easy to become involved in too many projects, and I have never properly learned to say no. As the only woman curator in the paleobiology department and one of relatively few across the museum in the 1980s, I was called upon to represent "diversity" on many occasions. To this day, women at the museum feel such conflicting demands on their time and energy, as we continue to be more willing to step up and serve the community than many of our male peers.

My life as an early career paleontologist in the 1970s and 1980s was, overall, a wonderful time of professional and personal growth, both in the field and in the museum. I know that I was lucky to have a positive experience throughout those years, and that it was largely because of the many male colleagues who treated me as an equal and provided opportunities along the way. Though it certainly was not always smooth sailing on the gender bias front, I was able to wall off those experiences and focus on what to me was most important and fulfilling: building collaborations and friendships with wonderful people and pursuing knowledge as a museum scientist.

Anna K. ("Kay") Behrensmeyer, Ph.D., is a curator of vertebrate paleontology and senior scientist at the Smithsonian's National Museum of Natural History, Washington, D.C.

Jenna Kaempfer

Ph.D. candidate in geochemistry and tectonics, University of Illinois at Urbana-Champaign

Dr. Penny Higgins

Vertebrate paleontologist and geochemist, University of Rochester

Dr. Lisa White

Micropaleontologist and science educator, University of California Museum of Paleontology

10

From Microfossils to Museums: Reflections on My Journey as an Earth Scientist

Lisa White

I often reflect on a field trip to southern Utah with urban high school students recruited from San Francisco, New Orleans, and El Paso in 2010. While looking at a particularly spectacular outcrop, one of the students turned to me and asked, "Is this real?" I was taken aback—after all the pre-trip preparation, the introduction to basic geological concepts and highlights of how landscapes form, this student asks if the rock features are real?! After asking the student what he meant by the question, I realized that little in his education prepared him to describe or understand the natural world or provided opportunities for outdoor experience. That student's question and my own never-ending excitement about teaching, learning, and communicating about geoscience, particularly in field settings, drives much of the work I do today.

My commitment to including a greater diversity of participants in Earth science stems not only from my own experiences as an African American woman geoscientist but also from a firm belief that authentic exposure to science at critical junctures in the training and mentoring of students triggers the kind of drive and lifelong dedication necessary to pursue science as a career. Many of my geology and paleontology peers cite experiences in their youth that eventually led them to a professional path in geoscience: never outgrowing a love of dinosaurs, family trips to national parks, school visits to natural history museums, and/or memorable hands-on science classroom experiences. Then there are individuals like me who discovered earth science later in our educational careers and, after exposure to a number of pivotal field and laboratory experiences, couldn't think of any other profession we would rather pursue.

Drawn more to the arts than the sciences while growing up in San Francisco, I majored in photography as an undergraduate student at San Francisco State University. Enamored of landscape photography, I had

ambitions to be an African American female Ansel Adams. As I became more curious about how landscapes physically formed, I took a geology class for a general education requirement and was fascinated by all things Earth. That initial course, and subsequent courses in my newfound major in geology, rekindled a love of science I had enjoyed in my youth. In San Francisco, I grew up near the California Academy of Sciences, where I was fascinated by the rock and mineral displays; perhaps my fascination with geoscience really started at the natural history museum. I had the good fortune to intern at the U.S. Geological Survey (USGS) while an undergraduate, and there I received mentoring from a number of female geologists and met my first female paleontologist.

Many of these women were more experienced than their male counterparts but were not promoted to the same level of leadership. Witnessing their professional struggles as women in the geosciences in the 1980s and the ways they forged ahead, managing and overcoming obstacles to become project leads, made me even more motivated to continue my career training in geoscience. Witnessing sexual harassment and the impact of unconscious biases shook whatever naive views I held about workplace equity at the time. It also cemented my commitment to supporting other women and underrepresented minorities in geoscience.

After earning a bachelor's degree in geology from San Francisco State

Dr. Lisa White leads the way through the expansive collections at the University of California's Museum of Paleontology (UCMP).

A closer look at some of the specimens archived at the UCMP

and a Ph.D. in earth sciences from the University of California at Santa Cruz, I was recruited by San Francisco State in 1990 for a faculty position in the Department of Geosciences. Teaching classes in paleontology, historical geology, and oceanography while directing research programs in micropaleontology, I frequently incorporated my training as a scientist with strategies to more effectively communicate and demonstrate geoscience concepts to diverse audiences. These included using local or place-based examples of geological change and drawing from environmental science and environmental justice themes to make geoscience concepts more relevant to students. Midway through my twenty-two-year faculty career at San Francisco State, I received the first of several National Science Foundation (NSF) grants to broaden the participation of underrepresented high school students in geoscience. Starting with the Reaching Out to Communities and Kids with Science in San Francisco (SF-ROCKS) program and continuing with the Minority Education Through Traveling and Learning in the Sciences (METALS) program, I implemented approaches common in successful broadening participation programs: field excursions within supportive learning environments, effective mentoring and peer interactions, and earth science instruction made relevant to the lives of diverse populations. I was intent on replicating some of the field experiences I'd had while interning at the USGS as an undergraduate student, going with professionals to faraway places from my perspective at the time (Alaska and northern New Mexico) to map mineral and water resources and better define geological hazards.

A desire to continue connecting and engaging underrepresented communities with science, particularly the paleontological sciences, led me to the University of California Museum of Paleontology in 2012; as the director of education and outreach, I manage the museum's education and public programs, including several award-winning web resources on evolution and the nature and process of science. A number of newly awarded NSF collaborative grants that I direct support diversity in geoscience field instruction, improve biodiversity literacy in undergraduate education, and train graduate students in science communication in ways that prepare professional scientists to teach and guide the next generation of geoscientists.

Lisa White, Ph.D., is a micropaleontologist and Director of Education and Outreach at the University of California Museum of Paleontology.

Dr. Emily Orzechow and Dr. Caitlin Boas

Finnegan Lab, University of California, Berkeley

Tripti Bhattacharya
Byrne Lab, University of California, Berkeley

Camilla Souto and Lucy Chang

Marshall Lab, University of California, Berkeley

Ashley Hall
Marketing coordinator and educator, Cleveland Museum of Natural History

11

Can You Be a Paleontologist Without a Ph.D.? (The Answer Is Yes)

Ashley Hall

"Did you marry into paleontology or are you actually pursuing it?"

This was an actual question posed to me—not by an older man, but by a woman co-worker my age. Stunned, I wasn't sure how to respond. At the time, I was working as a museum educator and assistant curator in a small paleontological museum, but because I didn't run my own lab or hold a Ph.D., in her eyes I had only "married into" paleontology.

Paleontology isn't a field that anyone is seeking to "marry into," though. No one makes enough money in our field that any woman would want to be a paleontology "trophy wife"—as glamorous as that sounds. Encountering bias in those of the same gender, as in my situation, is not uncommon, but is still surprising (and insulting) when it happens.

Women in the geosciences have had a long, difficult road for the last few hundred years. From the nineteenth century, when a young Mary Anning first set foot on the Jurassic coast of southern England and found her very first fossil, women have been documented as being involved in paleontology. Anning was a young, groundbreaking, trowel-blazing woman whose accomplishments, much like her fossils, have only now been written in stone. She was not accepted into the Geological Society of London because she was a woman. It is only now, in the twenty-first century, that Anning has finally been given the honor and respect she deserves; it is only now that she is being called a paleontologist; it is only now, after 200 years, that women in the geosciences hold her in the highest possible regard for facing the sexism and criticism that many women still endure. I know I do.

My personal journey has not been easy. From the time I was four years old, I have dreamed of being a paleontologist. Even to this day, I question if my career path was "the right one." Sometimes it's more organic than that. When I was a college student at Indiana University Bloomington, math was my least favorite subject and I was consistent, if nothing else, at barely passing. Later in life, I believe that my failures were due to undiagnosed dyscalculia or "math dyslexia." Because of this, I majored in anthropology instead of geology, as it required only a minimum of one math class. Determined to make my own path to paleontology, I focused on zooarchaeology and spent four years processing the remains of bison that had been discovered at protohistoric habitation sites under the supervision of an amazingly supportive professor, Dr. Laura Schieber. From this training, I was able to apply my knowledge of vertebrate anatomy to paleontology. Determined to be a paleontologist, I moved to California and found two careers: one as a museum educator at the Natural History Museum of Los Angeles County and the other as assistant curator at the Raymond M. Alf Museum of Paleontology.

I don't have a Ph.D. Does that make me any less of a paleontologist? What makes someone a paleontologist, geologist, or any other scientist is their contributions to the field. In my career, I have educated thousands of students in natural history museums and catalogued and curated thousands of fossils that contribute to our scientific knowledge of the past. I have spent hours in the badlands of Montana with my face inches from the clay, looking for 75 million-year-old dinosaur eggshells; over a decade educating the public about misconceptions of dinosaurs; and years of my life on social media platforms communicating about science to the public. I am a paleontologist.

Paleontologists pursue paleontology because of their passion for studying ancient life, and there are many pathways to a successful and meaningful career. If you're thinking about paleontology, consider your options. While a Ph.D. may be great for some, it's not for everyone. Ask for advice from mentors. Reach out to scientists on social media. We have more resources at our fingertips than every generation that's come before us. Do what's right for YOU and you will succeed!

Ashley Hall is the marketing coordinator at the Nature Center at Shaker Lakes.

Liz Ferrer and Sarah ElShafie

Padian Lab, University of California, Berkeley

ReBecca Hunt-Foster

Vertebrate paleontologist, Bureau of Land Management, Utah

12

The Path Is Not Always Straight

ReBecca Hunt-Foster

I have known that I wanted to be a paleontologist since I was thirteen years old. I also knew that I was interested in working at a museum or a national park. The only clear path from point A to point E that I knew of was to attend college and obtain a Ph.D. At that time, I was unaware that there were other options, and that deviating from these expected paths could still lead me toward becoming a paleontologist.

Although I was a decent student in high school and had supportive science and art teachers, my algebra teacher knew I was struggling and told me: "Not everyone is meant to go to college." I kept trying, graduated, and continued on to college, where I still struggled with some classes. Jack Horner, a well-known professor of paleontology who has dyslexia, suggested that I get tested for a learning disability, after which I found out that I had a severe math learning disability and dyslexia. With my diagnosis came tools to help with the way my brain works, something I had never had before.

I had many opportunities to explore potential career paths during college: working at a campus museum, interning for Alaska State Parks, assisting an anthropology professor, and attending academic conferences. Each experience helped me to better clarify what my ideal job was. I continued with my education by working toward my master's degree in geology (with an emphasis in vertebrate paleontology), with plans to follow a "traditional path" and obtain a Ph.D. My GRE scores were not great, and I was not able to get into any programs. I was told by my advisor that not everyone should get a Ph.D., which left me feeling lost, unsupported, and depressed.

No one had ever told me there could be other viable career paths to realizing my goals, and I had witnessed academic snobbery where some looked down upon anyone who had not earned a doctorate. And I was unaware that some of the more talented, sharp, and gracious paleontologists I knew did not have a Ph.D. I started to apply for internships and jobs and landed both in the first few weeks of looking. I turned a job down in favor of an internship at Glacier National Park, and, at the end of the summer, began the job search again. I had three interviews and secured a grant-funded position at Augustana College (Rock Island, Illinois) to prepare fossils from Antarctica. My supervisor, Dr. Bill Hammer, gave me the flexibility and support to grow professionally. I worked at Augustana for a wonderful three years before moving to the Museums of Western Colorado's Dinosaur Journey Museum.

My first two years were spent as field coordinator, and I had the opportunity to spend much of the summer digging for late Jurassic-aged dinosaur fossils with members of the public. I then transitioned to working as paleontology collections manager for the next four years and was able to use my organizational skills. Most of the tasks were perfect for the way my brain works: handling fossils, cataloguing them, and assisting visiting researchers with their needs. During this time I also worked as a consultant, conducting field projects for energy companies and writing reports for the Bureau of Land Management (BLM) and National Park Service, which helped build my skill set. At the beginning of 2013, I began my job as the district paleontologist for the BLM in southeastern Utah, where I worked until 2018. My work included everything from education and outreach (developing new trails, signage, and programs) to doing project clearances for other resources. I also facilitated paleontology research by helping to process permits and organize clearances for excavations. Some of the biggest rewards came from mentoring interns and helping our local communities and visitors recognize the scientific value of the local paleontological history through educational events, volunteering opportunities, and avocational groups, such as the Utah Friends of Paleontology. Managing fossils on public lands using science-based decisions made me feel that finally, I had a chance to define my professional goals and professional identity. I was honored to be selected as the new paleontologist at Dinosaur National Monument during the summer of 2018—recognition of my achievements that were all accomplished without a Ph.D.

As I have moved through my career, it has become clear that one does not need a Ph.D. in order to be successful—there are many potential paths. The details are in the climb: you need to be persistent and ambitious, contribute to science, make your own decisions, and not let others decide your fate.

ReBecca Hunt-Foster is monument paleontologist at Dinosaur National Monument, located in northeastern Utah and northwestern Colorado.

Dr. Kelli Trujillo

Vertebrate paleontologist and stratigrapher, Uinta Paleontological Associates

Dr. Karen Chin

Paleoecologist, University of Colorado Boulder

13

A Less Traveled and More Meandering Path

Karen Chin

Being a paleontologist has been a tremendously rewarding career, but my path to becoming a professor was a circuitous one. As bright-eyed children, we expect that our life journey will be straight, easy, and always positive. However, life is complicated.

I was a good student in high school. When I was an undergraduate, however, health issues wreaked havoc on my grades, and I went from being in the ninety-eighth percentile in high school to being on academic probation at my university. I managed to graduate, but my self-esteem plummeted. Suddenly, life was much more difficult, and my path forward was cloudy. I was beginning to learn that life can turn out quite differently than we imagine it will.

Yet taking an unexpected detour is not necessarily a bad thing. During the next phase of my life, I happily focused on my work as a seasonal National Park Service interpreter, leading nature walks and teaching park visitors about the natural world. I adored this job; I watched wild grizzlies and mountain goats, led campfire songs, tramped in alpine meadows, and saved a child from drowning. Yet I had to put food on my table when I was not in the parks, so I scrambled to find a variety of temp jobs that ranged from selling clothes and working at a ski resort to doing housekeeping.

Living a seminomadic life was never dull as I switched between work in the parks and other jobs. Nevertheless, I eventually knew that it was time to change course and aim for year-round employment. After serious self-reflection, I decided to pursue a museum career and went back to school. Even though I had once vowed never to go back to school, I was surprised to find that I enjoyed graduate studies. Still, I found that I was not particularly passionate about my research on native grasses.

It took a serendipitous event to turn me in the direction of paleontology. While working on my master's degree, I took a job preparing dinosaur bones for paleontologist Jack Horner at the Museum of the Rockies, Montana. Once I was introduced to paleontology, I couldn't learn enough about it. I was fascinated that fossil evidence from ancient organisms could be preserved despite the passage of millions of years, and realized that this interdisciplinary science really fit my broad interests.

Therefore, I gave up my seasonal work in the parks and made another course change, entering a doctoral program in paleontology. The pathway to a career as a professor has been more focused, but it has not been easy. Indeed, it has been an uphill climb with periods of stress and joy. Like most graduate students, I wondered if I would ever finish my dissertation. And it was surprising to discover that being a professor was also stressful, though in different ways. But the thing about journeys is that they are usually enriched by traveling companions; I owe a lot to family, my graduate advisor, and other generous mentors who had faith in my abilities and offered invaluable opportunities that helped me navigate this path.

I share my story to remind others that unconventional life journeys can be just as rewarding as more traditional ones. I used to deeply regret that I had not taken a more direct route to becoming a scientist. I have finally realized, however, that my atypical path was the right one for me. Being a park naturalist gave me unique perspectives on my studies of ancient ecosystems because I had the opportunity to learn about things as disparate as seasonal changes in moose diets, ladybug population explosions, and blue fungi in pine wood. Indeed, I now study fossilized feces because I found out that animal scat can tell us much about ecosystems. Moreover, the personal struggles, health issues, time spent exploring, and diverse jobs of my "nonacademic" years shaped who I am as a person and as a scientist. I am not a better or worse scientist for it; I just took a different path.

Karen Chin, Ph.D., is an associate professor in the Department of Geological Sciences and curator of paleontology at the Museum of Natural History at the University of Colorado Boulder.

Dr. Cindy Looy
Paleobotanist, University of California, Berkeley

Natalia Villavicencio, Emily Lindsey, and Allison Stegner

Barnosky Lab, University of California, Berkeley

Jenna Judge and Rosemary Romaro

Lindberg Lab, University of California, Berkeley

Erica Clites, M.S.
Invertebrate paleontologist, University of California Museum of Paleontology

Dr. Patricia Holroyd
Vertebrate paleontologist, University of California Museum of Paleontology

Women

No experience of womanhood is ever singular or universal; all women must negotiate their positionality in regard to race, gender, sexuality, class, religion, ability, and/or nationality.

Amy K. Guenther

Ingrid K. Lundeen and Amy L. Atwater

University of Texas at Austin

14

Definition: Woman

Amy K. Guenther

Why even define "woman"?

It may seem unnecessary. Many people have never questioned and/or thought about what a woman is unless they or someone near them have had their identity as a woman questioned or been called a woman when they are not. Hegemonic society assumes a woman is a female with a vagina who menstruates once a month; can bear children; and can be a mother, wife, daughter, sister, or, more recently, a scientist, CEO, or even president. This implicit understanding of "woman" is encountered every day and has been used to organize many societal systems, from medicine to education. Yet "woman" is not so easily defined. The problem with relying on an unexamined conceptualization of "woman" is that it often tends to exclude those on the edges of what hegemonic society defines as "woman."

Indeed, for centuries feminists have questioned, critiqued, and redefined what it means to be a woman: from English writer and philosopher Mary Wollstonecraft's examination of women in her 1792 *A Vindication of the Rights of Woman*; to abolitionist and women's rights activist Sojourner Truth's 1851 speech "Ain't I a Woman?" addressing racism in the women's suffrage movement; to French philosopher Simone de Beauvoir's now famous statement that "one is not born but, rather, becomes a woman," in her 1949 *The Second Sex*; to legal scholar and critical race theorist Kimberlé Crenshaw's coining of the term "intersectionality" in 1989 to describe the interlocking oppressions of being black and a woman;[1] to trans women in the twenty-first century claiming their space as women.[2] These examples demonstrate that there is no one singular definition or experience of "woman."

Even today, we as a society continue to struggle with how to define "woman." Both the *Oxford Dictionary* and the *Collins English Dictionary* have recently faced controversies over what many perceive as sexist language associations and trans-exclusionary definitions of "woman."[3] While the publishers argue that dictionaries reflect words in their common usage, which is not necessarily how scholars or activists might define them, dictionary definitions of sex and gender terms do have real-world consequences.[4] Moreover, as the renewed controversy around Olympic champion Caster Semenya also reveals, the International Olympic Committee and the International Association of Athletic Federation have a long, problematic history of using medical testing to determine if a woman athlete meets certain biological standards of being female and thus is eligible to compete in women's athletics.[5] In Semenya's case and those of many other women of color outside of sports, the definition of "woman" also has much to do with race and racism.[6] So when a definition is left unexamined, hegemonic society still privileges and implicitly assumes "woman" to be biologically female, cis, white, and heterosexual. Therefore, in (re)examining definitions of "woman," *The Bearded Lady Project* attempts to define the term as broadly and inclusively as possible.

Woman (noun and an apposite noun)
We use the term "woman" as a type of gender category. Therefore, it is socially, historically, and culturally constructed. As such, "woman" has meant different things in different times and places. As a gender category, "woman" exists on a spectrum of masculinity and femininity, which are also socially constructed categories. "Woman" is not synonymous with "feminine" or "femininity." Likewise, masculinity is not antithetical to "woman." Many in the public use "woman" interchangeably with "female," but gender scholars and activists tend to reserve "female" for biological sex rather than gender categories.

We also use the term "woman" intersectionally, meaning that no experience of womanhood is ever singular or universal; all women must negotiate their positionality in regard to race, gender, sexuality, class, religion, ability, and/or nationality. Perhaps the most readily available example of the intersectionality of (at least) gender and race is the gender pay gap, wherein for every dollar a man earned in 2017, a woman earned 80 cents. Broken down further, however, for every dollar a white man earned, Latinas earned 53 cents, Native American and Alaska Native women earned 58 cents, black women earned 61 cents, white women earned 77 cents, and Asian women earned 85 cents. Yet even these categories can be further broken down and complicated by factors such as sexuality, motherhood, disability, and age, among others.[7]

We acknowledge that the term "woman" is often used in medical and scientific research and public discourse to denote the biological sex of people born with what are generally understood as female reproductive organs (vagina, ovaries, no Y chromosome, etc.); however, we problematize this definition as inadequate and exclusionary.

- It does not fully encompass the existence of intersex women (people who possess male and female chromosomes, hormones, genital, and/or sex organs and tissues), trans women, and cis women with elevated androgens (male hormones).

- A biological definition also incorrectly includes people with female reproductive organs who are not women. We understand that not all women have vaginas and not all people with vaginas are women.

- Additionally, it brings into question the "womanhood" of cis women who never menstruate, never become pregnant, have had a double mastectomy, and/or have had a hysterectomy. If being a woman is based solely on biology, an argument often made by both the far right and the more liberal trans-exclusionary radical feminists (TERFs), then does changing these biological markers change a cis woman's identity as a woman?

- Moreover, biological definitions have been dangerously aligned with, if not used as outright justifications for invasive medical surveillance of and violence against women of color and women deemed "sexually deviant."[8] For example, nineteenth- and early twentieth-century scientists and doctors frequently studied (often enslaved) black women's genitalia and bodies, with questionable levels of consent, for anatomical anomalies to explain what the doctors perceived as heightened sexuality and to justify biological differentiation between races.[9] Forced sterilizations have also been overwhelmingly carried out against women of color in the United States in a messy confluence of racism, misogyny, ableism, and classism. Such an extreme biological alteration is meant as both a form of eugenics and a means of curbing so-called deviant behaviors that exceed hegemonic gender and sexuality norms.[10]

Biological definitions are still being used to justify discrimination against women (when "woman" is meant as female assigned at birth). As recently as 2018, CERN scientist Professor Alessandro Strumia, a member of one of the most well-known scientific organizations in the world, gave a talk on gender and physics in which he suggested that men's brains are better suited to physics than women's.[11]

For all of these reasons, "woman" is a term inadequate to capture the complexities of gender identity when it is simply juxtaposed against the dichotomous term "man"; however, it is still the best term we have (right now) to describe and study gender inequities in society and the sciences until researchers widen the gender demographics studied.[12]

We admit that by necessity our definition is incomplete and ongoing.

Amy K. Guenther (she/her) is a freelance scholar, dramaturg, and teacher in Austin, Texas. She has a Ph.D. in theater history, literature, and criticism with an emphasis in performance as public practice from the University of Texas at Austin.

---

1. While certainly not the first to articulate such interlocking oppressions, she was the first to use the term "intersectionality" to describe it. Today intersectionality is used more colloquially to mean the intersections of gender, race, sexuality, class, and other factors. "Kimberlé Crenshaw on Intersectionality, More than Two Decades Later," Columbia Law School, accessed October 2, 2019, https://www.law.columbia.edu/pt-br/news/2017/06/kimberle-crenshaw-intersectionality.
2. This is not to say that trans people have not always existed but to observe that, more and more, language is being created to articulate their experiences in this century as trans rights also gain more and more visibility in public discourse. The inclusion of trans women has been one of the primary arguments in redefining "woman" in the twenty-first century. For a brief history, see Carol Hay, "Who Counts as a Woman?" *The New York Times*, April 1, 2019, https://www.nytimes.com/2019/04/01/opinion/trans-women-feminism.html.
3. Nora Caplan-Bricker, "Is It Time to Change the Definition of 'Woman'?" Slate, September 29, 2017, https://slate.com/human-interest/2017/09/why-a-controversial-definitionof-the-word-woman-doesnt-necessarily-mean-the-dictionary-is-sexist.html; Allison Flood, "Sexism Row Prompts Oxford Dictionaries to Review Language Used in Definitions," *The Guardian*, January 25, 2016, https://www.theguardian.com/books/2016/jan/25/oxford-dictionary-review-sexist-language-rabid-feminist-gender; Kathryn Madden, "The Dictionary Definition of 'Woman' Needs to Change," *Marie Claire*, July 5, 2019, https://www.marieclaire.com.au/dictionary-sexism-woman.
4. Caplan-Bricker cites three cases that use dictionary definitions and common usage arguments to decide court cases: *Chambers v. Ormiston* 2007 in the Rhode Island Supreme Court and *re Estate of Gardiner* 2002 in the Kansas Supreme Court, which was then used in *re Application for Marriage License for Nash* 2003 in an Ohio court of appeals. Caplan-Bricker, "Is It Time."
5. Tellingly, sex verification testing is not done for men's athletics. Michelle Garcia, "Our Cover Star, Caster Semenya: The Athlete in the Fight for Her Life," *Out*, July 23, 2019, https://www.out.com/sports/2019/7/23/our-cover-star-caster-semenya-athletefight-her-life.
6. Anna North, "'I am a Woman and I am Fast': What Caster Semenya's Story Says About Gender and Race in Sports," Vox, May 3, 2019, https://www.vox.com/identities/2019/5/3/18526723/caster-semenya-800-gender-race-intersex-athletes. Also see germinal works in women of color feminisms such as Cherríe Moraga and Gloria Anzaldúa, eds., *This Bridge Called My Back: Writings by Radical Women of Color*, 2nd ed. (Latham, NY: Kitchen Table, Women of Color Press, 1983); Combahee River Collective, "Combahee River Collective Statement," in *Home Girls: A Black*

Feminist Anthology, ed. Barbara Smith (New Brunswick, NJ: Rutgers University Press, 1983): 264–274; and bell hooks, *Ain't I a Woman: Black Women and Feminism* (Boston: South End Press, 1981).

7. Kevin Miller, Deborah J. Vagins, et al., *The Simple Truth About the Gender Pay Gap* (Washington, DC: American Association of University Women, 2018), 9–18, https://www.aauw.org/aauw_check/pdf_download/show_pdf.php?file=The_Simple_Truth.

8. Perceived sex variations also have a history of being used to explain the existence of lesbians. Jennifer Terry, "Lesbians Under the Medical Gaze: Scientists Search for Remarkable Differences," *The Journal of Sex Research* 27, no. 3 (1990): 317–339.

9. Siobhan Somerville, "Scientific Racism and the Emergence of the Homosexual Body," *Journal of the History of Sexuality* 5, no. 2 (1994): 243–266.

10. Lisa Ko, "Unwanted Sterilization and Eugenics Programs in the United States," *Independent Lens*, January 16, 2019, http://www.pbs.org/independentlens/blog/unwantedsterilization-and-eugenics-programs-in-the-united-states/.

11. Pallab Ghosh, "Cern Scientist: 'Physics Built by Men—Not by Invitation,'" BBC News, October 1, 2018, https://www.bbc.com/news/world-europe-45703700; Rafi Letzter, "A Physicist Said Women's Brains Make Them Worse at Physics—Experts Say That's 'Laughable,'" LiveScience, October 2, 2018, https://www.livescience.com/63730-physicistsays-women-bad-at-physics.html.

12. While writing this essay, we came across the concept of womxn but have chosen not to include it for several reasons. "Womxn" is a relatively recent term and, as such, there is still much disagreement on what it means and whom it excludes and includes. Many use the term to specifically include trans women (and some nonbinary people) and women of color as womxn. Both trans women and women of color have contentious histories of inclusion in the category of woman, as discussed in this essay. Most pertinent to our purposes, however, is the fact that womxn is extremely controversial in the trans community because it tends to exclude trans women (and by extension women of color) from the category of woman, thus reifying "woman" to always implicitly mean white and cis. This is by no means a universal understanding of womxn, however, and it is increasingly being used on U.S. college campuses. For further reading: Breena Kerr, "What Do Women Want?," *The New York Times*, March 14, 2019, https://www.nytimes.com/2019/03/14/style/womxn.html?searchResultPosition=1; Alex Regan, "Should Women Be Spelt Womxn?," BBC News, October 10, 2018, https://www.bbc.com/news/uk-45810709; Luna Merbruja, "3 Common Feminist Phrases That (Unintentionally) Marginalize Trans Women," Everyday Feminism, May 12, 2015, https://everydayfeminism.com/2015/05/feministphrases-marginalize-trans-women/; Asia Key, "Woman, Womyn, Womxn: Students Learn About Intersectionality in Womanhood," *The Standard* (Missouri State University), March 27, 2017, http://www.the-standard.org/news/woman-womyn-womxnstudents-learn-about-intersectionality-in-womanhood/article_c6644a10-1351-11e7-914d-3f1208464c1e.html.

Tesla Monson, Whitney B. Reiner, and Dr. Leslea J. Hlusko

Hlusko Lab, University of California, Berkeley

15

The Moments When I Am Not a Woman

Leslea J. Hlusko

My first paleontological field experience was in Kenya at the Koobi Fora Field School in 1991. Most of my memories are cringeworthy, as my twenty-year-old attention was focused more on what people thought of me than on the amazing environment. But there was a critical moment that tilted the direction of my career aspirations toward fieldwork. An instructor took us on a geology walk to learn the depositional and tectonic history of the area. I, in my dusty clothes and unwashed hair, lost track of my interpersonal concerns as I jumped from boulder to boulder, absorbing the idea that these were once living algae mats, since frozen in time.

This was a first for me—to think about a scientific fact without my life experiences and personal concerns being a part of the framework in which I considered it. Sitting in a classroom, I was never free from the people around me, always worried about what they were thinking of my clothes or unsure what questions to ask. But the time depth of those rocks was so much larger than the life of one person—of me—that I felt irrelevant, without self-consciousness. I felt free.

The rewards of fieldwork are many. But it is these moments, when I lose myself in geological time, that I love the most. Over the last twenty or so years of fieldwork in Ethiopia, Kenya, and Tanzania, I have sat on the edge of gorges and at the top of hills overlooking rocks that encapsulate millions of years of evolution, many of them containing the fossil remains of humanity's ancestors. I am not a woman at these moments. I am a human. I am a hominid. I am an animal feeling the direction of the wind on my face and the smell of the Earth in my chest. I am one brief moment in the evolution of life. I am alive. I am nothing. I am a part of everything.

But these moments are fleeting.

Primate skull display in the Hlusko Lab

I must not contemplate too long, lest someone ask me (yet again) if I am tired from the physical work. I must walk at the front of the crew and carry one of the heavier loads of equipment. My every action, decision, movement is judged for signs of weakness. "Why is this woman here, without a husband, and apparently in charge?"

No wonder Mary Leakey—a British paleontologist who discovered the first fossil believed to be the ancestor of apes and humans in 1948—is rumored to have had an affection for whiskey. It is exhausting to constantly be monitored for inferiority. This is a man's world. But here I am—one fleeting and irrelevant configuration of biology in the expanse of evolutionary time. Here I am, for a moment.

Leslea J. Hlusko, Ph.D., is professor of integrative biology and associate director of the Human Evolution Research Center, University of California, Berkeley.

Tara Smiley
Paleoecologist, Oregon State University

Katharine Loughney

Ph.D. candidate in vertebrate paleontology, University of Michigan

Dr. Catherine Badgley

Paleoecologist, University of Michigan

Rainbow Basin field crew
Mojave Desert, California

Dr. Lisa Boucher

Paleobotanist, University of Texas at Austin

Dr. Andrea D. Hawkes

Micropaleontologist, University of North Carolina Wilmington

16

My Love-Hate Relationship… with Waders

Andrea D. Hawkes

Not all marshes are created equal. For example, marshes that have a lot of sand or are very peaty are relatively easy to walk across. It is the less vegetated, silty, clayey, typically lower parts of the marsh (called the low marsh and tidal flat) where walking can be the most difficult. In fact, I wouldn't call it walking, I'd call it trudging.

Trudging is the act of walking on a marsh surface whereby the suction of the surface challenges the strength of your quads and your physical and mental stability. If you work in marshes (salty or fresh ones), you have some understanding of what I am talking about.

As I am a coastal geologist who studies natural hazards (earthquakes, tsunamis, storms, rising sea levels), much of my research requires trudging in coastal marshes, and I have been lucky enough to do this in a number of interesting locations worldwide (Indonesia, Argentina, South Africa, etc.). In addition to all the required field notes a good scientist takes, I try to include comments about the marsh trudginess. I write things like: "robust surface," "soupy but no suction," or "death trap, lost two students." (I did once have two students end up barefoot. They lived, but their boots were sacrificed to the marsh deities forever.) I make jokes, but people have actually died from getting stuck in marshes, especially in areas with high tidal ranges where they just can't get free before the high tide drowns them. It is a thing of nightmares.

This brings me to the topic of waders and my love-hate relationship with them. I love them because they keep me dry and clean, and my existing pair have this awesome kangaroo pocket in the bibbed part of the overalls, which is great for a field book, a pencil, electrical tape, and a granola bar; I hate them because they never fit and they can kill you (more about this later)! The durable ones are generally men's sizes, so they are way too long (and wide) for my totally average, 5-foot-3-inch female frame, which means

that when they're pulled up as high as possible, the top of the bib lands at my shoulders—and still they are too long.

This matters, because if you are trudging you have to hold the excess leg fabric in your hands so that when you walk the boot doesn't stay in the marsh while your foot comes out of the boot (this generally leads to tripping/falling). Why not get a boot size that fits? It is my men's shoe size! (Waders are sized by shoe size, so size 4 men's equals size 6 women's.) Who are these men's size 4 waders made for, anyway? They seem to be designed for someone who is at least 6 feet tall (37-inch inseam), which is a rare aspect ratio given that most people who are 6 feet tall don't wear men's size 4 shoes (or vice versa).

Because my hands are busy holding up pants legs to ensure that the boots stay on my feet, carrying field equipment is complicated and more difficult than necessary. And if you don't carry some gear, you feel like you are not doing your part; it's a vicious physical and emotional cycle.

If the boots were more secure around the ankles, it wouldn't be as much of a problem. It has been suggested to me, "Duct tape the boot tighter at the ankle so your feet can't come out." I thought, *Does this person want to kill me?* Because this is a big no-no. If I did what they suggested and I happened

Isabel Hong trudges through the deep mud at her field site in Raymond, Washington.

to fall into a channel or river, my waders would fill with water, I might not be able to get them off, and I could drown. Again, the stuff of nightmares.

In the grand scheme of things I know this isn't a big deal, but it does make my job a little more difficult. On a recent field trip I had so many bags for field equipment that I didn't have space to pack waders and hadn't thought to mail them ahead. (Yes, I have done this, for the exact reason I am about to describe.) I thought, *Oh, for sure, by now store x or y (major outdoor apparel outlets) will have my size either in store or online.* Except that totally wasn't/isn't the case. Unlike many of my colleagues, who can pop into a store to replace their busted waders while out on fieldwork, I would have to go without—and did. To reiterate: although this isn't a big deal, sometimes all the little things added together just make my job that much harder. Simply put: Why can't there be waders that fit?

My first experience in a marsh was in Nova Scotia during my undergraduate studies, and I was hooked after I learned a few valuable lessons for marsh trudging:

1. Never jump and land on both feet! You always need one leg free in order to get yourself unstuck.
2. Walk on or very near vegetation. The root structure holds up the marsh surface. However, this isn't always possible in tidal flats, so be careful.
3. Always do fieldwork with more than one person. This is true of most fieldwork. If you do plan to go solo, at the very least, let someone know where you are going and check in. Another clever trick is to carry something that will help your trudging experience. For example, because I am either coring the marsh or taking surface samples, I often carry a gouge core (think long steel *T*) or level pole (one-inch-thick telescoping pole) that I can also use to test out the trudginess or stability of the marsh surface. This can also be used to help me jump creeks (think pole vaulting), and if I do get stuck, it provides the leverage and hard surface area I can use to get free.

It has been a long while since I have been stuck in a marsh, but every year I see a new flock of students navigate marsh trudging and, as funny as it always is, I just think, *If only the waders fit!*

Andrea D. Hawkes, Ph.D., is an associate professor in the Department of Earth and Ocean Sciences and Center for Marine Science at the University of North Carolina Wilmington.

Isabel Hong
Ph.D. candidate in paleoseismology, Rutgers University

Dr. Tina Dura
Paleoseismologist and micropaleontologist, Rutgers University

Dr. Patricia H. Kelley

Invertebrate paleontologist, University of North Carolina Wilmington

17

The Balancing Act

Patricia H. Kelley

I was pregnant and scared—and not for the usual reasons.

I was two years into my tenure-track position at the University of Mississippi. My husband and I had decided to start a family, and I was worried that having babies would jeopardize my tenure bid. It was 1982. Up to this point, no mother had ever received tenure in the School of Engineering, because I was the first and only female faculty member. I was an oddball. I was not only a paleontologist who studied the evolution of clams and snails (in a geology and geological engineering department) but also surrounded by male engineers who were clueless about what paleontologists—and mothers—do.

When I couldn't hide the pregnancy anymore, I nervously confessed to department colleagues. One of them sighed in frustration, because he thought it meant he would have to go back to teaching a course I had been hired to handle. In shock, I realized that they assumed motherhood would end my career.

It could have. There was no stopping the tenure clock back then, and the Family and Medical Leave Act had not been passed yet. I had no maternity leave or on-campus daycare. Complicating things further, to allow my "trailing spouse" to pursue his career as a Presbyterian minister, I had to drive an hour and a half each way to and from the university, because we were expected to live in the parsonage next to the church.

We timed the first pregnancy well; Timothy was born in May, and I had the summer with him before classes started. I remember hand writing (it was 1982, after all) a manuscript on mollusk evolution as my son sat propped up in his baby seat on the couch next to me that summer.

A few years later, I became pregnant again. I told my department chair in March that I was expecting a baby in August. He responded by suggesting

I postpone the birth until October, when it would be more convenient for a colleague to take over my classes. I told him I didn't think I could delay childbirth for two months after the due date!

I scheduled a C-section to ensure that the birth would be as early as possible, so I wouldn't miss too much class while waiting for my doctor's permission to return to work. My (female) grad student taught the first two weeks of classes for me, and I was back teaching full time when Katherine was three weeks old. It was brutal, but I knew I couldn't let motherhood slow my productivity. Three years after Timothy was born, I got tenure, and four years later I was promoted to full professor.

One might wonder how I managed to surmount these tenure and promotion hurdles while raising a family and supporting my husband in his career. (Being married to a pastor comes with its own set of church responsibilities!) My husband had flexibility to schedule his church meetings when I could be home with the kids or when parishioners could babysit. I worked from home on days when I didn't have classes or meetings, grading papers while keeping one eye on the kids as they played in the church playground. The kids and I also bargained: "Let Mama do her very important college work and then we'll play pretend."

Academia post tenure has its own challenges. My workload remained intense; as the sole woman, I was in high demand for committee membership and was given administrative assignments (associate dean of engineering; acting associate vice chancellor for academic affairs). Managing all of these responsibilities required that our family continue to work together well.

We did schoolwork together. Even in high school, Katherine sat next to me on the bed doing calculus while I prepped for classes. And everyone pitched in to get the household chores done. When Timothy left for college, his classmates planned to come home on weekends so their mothers could do their laundry. In contrast, Timothy rejoiced that he no longer had to do the whole family's laundry.

Administrative experience provided mobility, and after a stint at the National Science Foundation, I ended up as the chair of geology and geological engineering at the University of North Dakota—again the lone female faculty member in the engineering school. Then, halfway through my career, I moved to the University of North Carolina Wilmington as chair of a department in which I had women colleagues (plus the location provided easy access to my beloved Coastal Plain mollusk fossils).

In these leadership positions, I aimed to help early career women (and men) succeed by removing the academic roadblocks I had faced. I served on advisory councils and committees that developed policies and programs to improve the culture—including leave of absence policies—for faculty and graduate students, and mentoring programs. Within my department, I arranged teaching schedules to accommodate pregnancies and childcare; I nominated female faculty and students for awards; and through formal mentoring programs, student advising (more than half of the forty theses and dissertations I advised were by women), and informal interactions, I tried to be the mentor I had never had—especially to those dealing with work-life balance issues. Now in retirement, I'm on the board of directors of the Association for Women Geoscientists, promoting their mission to enhance women's participation in the geosciences.

The academic world is different today. In the United States, the Family and Medical Leave Act protects our jobs if we take leave for maternity-related disability or to care for a child after birth, adoption, or foster placement—though it lags in comparison to regulations in some other countries regarding benefits accorded new parents. Stop-the-clock policies now exist at most colleges and universities but vary greatly among institutions, and those who use them may be stigmatized. Sadly, gender-based biases still exist.

Based on my own challenges, and against these odds, I offer the following suggestions to others navigating academia—be it pretenure, post-tenure, or in non-tenure-track positions—while trying to maintain a personal life: learn the system and know your rights regarding reappointment, tenure, promotion, leave, and other pertinent policies; take advantage of any formal mentoring programs and seek trustworthy mentors and colleagues who understand the (personnel, publishing, funding) system you need to navigate and who can answer your questions; seek colleagues who have dealt with or are facing work-life balance issues and can support you in your struggles; make sure your closest companions (family, friends) honor the goals you have for yourself and are willing to help you achieve them; recognize that work-life balance includes finding joy in helping others achieve their goals, and do all you can to enable the success of others—be it family, students, colleagues, or (in my case) fans of fossil mollusks.

Patricia H. Kelley received her B.A. in geology from the College of Wooster and her A.M. and Ph.D. from Harvard University. A former president of the Paleontological Society and recipient of the 2014 United States Professor of the Year Award, she is currently professor emerita of geology at the University of North Carolina Wilmington.

Dr. Bonnie Jacobs

Paleobotanist, Southern Methodist University

18

Being Brave

Bonnie Jacobs

Every person who faces inequities brings to that challenge the attitudes and teachings of their upbringing, including broader societal pressures and expectations. Being sixty-six years old, I recognize that I grew up in a society in which girls were not expected to be geologists. Geology was, and to some people even now is, seen as a masculine pursuit. In my day I had to pick and choose my battles carefully. Would I stand up to my undergrad advisor in order to get equal time in the field with him, while telling him, no, I would not have sex with him? (I did tell him no.) Would I ignore comments such as: "Oh, you don't look like what I expected when your mother told me you were a geologist" (I let it go); "Girls shouldn't be geologists" (I spoke out); "She's cute, but she'll never finish a Ph.D." (Again, I let it go), and on and on?

These offenses seem small to me now compared with the bigger challenges of more subtle and manipulative discrimination, such as being ignored in a faculty meeting, being paid much less than male colleagues, or being denied review for promotion to full professor while the males are taken more seriously. These all happened to me as well, and not that long ago. Sometimes I was not particularly brave. I learned quickly that speaking out prompts reactions from those who do not want their motives questioned. Those reactions often entail additional dismissiveness, and what I really feared most was to be labeled as bitter, difficult, or simply as having unrealistic expectations for myself—which would further alienate me from my immediate social-professional habitat. I forced myself to be brave and spoke up when I felt wounded or cornered, but mostly kept my head down to move forward. I wanted to do science, and to be good at what I wanted, so much that I just kept getting back on that horse whenever I felt like a failure. I nearly quit academics altogether when my children were young, but realized that paleontology was the only work-related thing that really brought happiness and satisfaction. To return and try to get a tenure-track job after being an adjunct for seventeen years, much of that part-time, was truly terrifying.

Ellen Currano and I met in 2008, after which she was funded by a National Science Foundation postdoctoral fellowship to work with me in Ethiopia. We have worked together continuously since then and are good friends. So when she began this journey with *The Bearded Lady Project* before she had tenure, I feared for her. The project has a point common to such endeavors: women and men of science should be approached with equal respect. However, a unique and more pointed message is conveyed by this question: If you see a woman in the field with a beard on her face, why does she suddenly look like she belongs there? The black-and-white images of women looking like men challenge the thoughtful viewer to ask why he or she may think that way. Others might miss this point, which caused me to fear that people would laugh at the whole endeavor—perhaps even be derisive and think it silly. Would they be on the defensive and discount our views because they did not want to think about their own complicity in the problems that stem from having greater expectations of male than female scientists? My concerns about these potential negative reactions caused me to worry that Ellen in particular was risking her personal reputation and perhaps even damaging her academic future. She was brave, but was she headed for trouble?

The overwhelming positive support for and participation in the project by both early and late-career women came from an inspiring leap of faith at the outset that she (and the entire *Bearded Lady Project* crew) would do a great job. Truly, they have, and if there was backlash, it was minor.

My own participation in *The Bearded Lady Project* took a bit of nerve, but I was already in the safety zone of my career and wanted to say yes when asked to help. I had tenure and the confidence that comes from experience, and I no longer cared so much about what people thought of me professionally. There was little risk, and it was even fun! Only occasionally do I need to be brave in the academic world today, and think carefully about picking my battles. But I find myself incredibly angry, as are many other women my age. Now that there are enough women paleontologists who are older and accomplished and work in an environment where it's okay to be angry, it all seems to be coming out. I care as much as ever about the harmful nature of sexism, but I am motivated much more now by any ability I may have to make it better for the women coming up after me.

Bonnie Jacobs, Ph.D., is a professor in the Roy M. Huffington Department of Earth Sciences at Southern Methodist University, Dallas, Texas.

Dr. Victoria Hudspith

Wildfire paleontologist, University of Exeter

Belcher wildFIRE Lab

University of Exeter

Stacy New
Wildfire paleontologist, University of Exeter

Charlotte Gurr
Paleobiology student, University of Exeter

Dr. Claire Belcher
Wildfire paleontologist, University of Exeter

Hannah Simpson
Earth science undergraduate student, University College London

Jessica Lawrence Wujek

Ph.D. candidate in vertebrate paleontology, University of Southampton

Kate Acheson
Ph.D. candidate in vertebrate paleontology, University of Southampton

Sarah Baker
Ph.D. candidate in wildfire paleontology, University of Exeter

Gender

As I was faced with putting on the beard, I felt again as if I were being asked to support the notion of a world in which there are only men and women scientists, and I couldn't help but feel like there were entire groups of my students being left out of this conversation.

Sara B. Pruss

19

Definition: Gender

Amy K. Guenther

Because *The Bearded Lady Project* directly challenges and pokes fun at the gendered stereotypes of paleontologists, it is important to define the term "gender." Much like with the term "woman," most people may never have considered what gender is unless they have been accused of performing it incorrectly. Moreover, the definition from which we begin our work—that gender is a social construction, not biologically based—has only been developing since the mid-twentieth century. Many gender scholars and activists specifically attribute the term in its current usage to sexologist John Money's research beginning in the 1950s.[1] Today, the breadth of language surrounding gender identity and expression is ever changing and expanding, largely via social media commentary and discussion among queer communities. Even social media behemoth Facebook introduced 50+ different gender options in February 2014—evidence that the visibility of genders beyond man/male and woman/female is evolving.[2]

At the same time, gender reveal parties have become more common and at times extreme. These parties celebrate the assignment of a binary gender to babies via color-coded revelations—for example, blue for boys and pink for girls—based on the interpretation of genitalia on an ultrasound. Thus, these parties conflate gender with sex assigned at birth. Many, including the inventor of such parties, have begun to question the ethics of parents celebrating, often extravagantly, the assignment of a binary gender, with all of its incumbent stereotypes, to children before children have any innate sense of self.[3] The continued prevalence of these parties demonstrates, however, that many in the general public have never really thought about gender as a social construction or stopped to examine their own implicit assumptions about gender. Therefore, it is important for *The Bearded Lady Project* to define our use of the word.

Gender (noun): the social, historical, and cultural expectations and norms around concepts such as masculine and feminine and how those concepts affect an individual's identity.

In general, U.S. society associates femininity with women and masculinity with men. These stereotypical associations create problems because no one is either all masculine or all feminine; we are all a combination of both, and no combination is inherently good or bad. One example of how these stereotypes create problems is that traditional society in the United States tends to accept some divergences, such as girls wearing pants and playing in the dirt, while expressing discomfort with others, such as boys wearing dresses and playing with dolls. This, of course, is also an example of a double standard. There is nothing morally or ethically wrong with girls or boys who eschew gendered stereotypes. Examples such as these, however, begin to reveal the social constructedness of gender.

Along with defining "gender" as a social construction, we acknowledge that "gender" describes how societal expectations and stereotypes interact with an individual's internal sense of gender identity, which they may or may not (be able to) express in their outward appearance or behavior. Traditional society tends to define gender as binary—either man or woman—but we recognize all genders.

- For example, "cis" and "trans" are two of the most commonly used adjectives to describe gender identity in the United States. A cis woman is a person whose gender identity (woman or female) matches their sex assigned at birth (female). A trans woman, however, is a person whose gender identity (woman or female) does not match their sex assigned at birth (male).

- "Trans" is an umbrella term, and not all trans people are binary, meaning that they transition to "man" or "woman." Some trans people are "nonbinary," meaning that they eschew the gender binary entirely and can be both male and female, neither, or a fluid combination of the two. Additionally, these definitions can be in flux. For example, two people who identify as "genderqueer" may have different definitions of what that means. Traditional society in the United States largely forces trans people to be more conscious of the social, cultural, and historical expectations around their gender than most cis people.

Even the imagery of the gender spectrum between masculine and feminine has more recently become insufficient for describing three-dimensional lived experiences, especially for trans identities. Such interactive images as "The Gender Unicorn"—which allows individuals to fill in levels of their gender identity, gender expression, sex assigned at birth, and physical and emotional attractions alongside a cartoon unicorn illustrating these concepts—have become more common in efforts to replace the language around binary genders (male and female) with more specific language.[4] They also help to more clearly delineate gender from sexuality, which are two distinct concepts; that is, one's gender does not determine one's sexuality.

Finally, because our definitions are situated in the United States, we must acknowledge that the stereotypes of gender, like other categories of identity, are inextricably linked to conceptions of race, heterosexuality, class, colonialism, and the assimilation to Western values.

- For example, the United States and other Western cultures have historically normalized gender as binary—either male or female—and imposed a binary gender system on colonized cultures. In the United States, trans and nonbinary genders are only now beginning to enter the popular consciousness. But many other cultures around the world have long identified more than two genders. Many Native American communities, for instance, have a third and sometimes fourth gender collectively called "two-spirit people."[5]

- Gender stereotypes are often centered in whiteness. For example, throughout U.S. history, black women have often been perceived as either too masculine or hypersexualized, outside the norms of proper (white) feminine behavior. This in turn has led to the questioning of their womanhood.[6] And Western standards of feminine beauty have historically been centered in whiteness: long, silky hair, often blonde; pale, alabaster skin; and thin noses and lips. Examples abound of this idealization of (white) feminine beauty throughout the canon of Western art.

- A worldview that sees heterosexuality as the norm often insists that men be masculine and women be feminine. For example, the perception and/or expression of femininity in men—whether in behavior, gesture, or career choice—has often been perceived as a sign of being gay, whatever their sexuality; conversely, society often stereotypes all gay men as feminine. These associations combine homophobia with toxic masculinity to create the fear that femininity in men is a sign of both weakness and being gay.

These, of course, are not the only examples of how gender intersects with other identity categories, complicating the concept of gender even further. Therefore, this essay is only a beginning on how *The Bearded Lady Project* defines gender. By necessity, these definitions are at once complicated and incomplete, because gender is always already in a process of becoming. That is, gender is performative.

Amy K. Guenther (she/her) is a freelance scholar, dramaturg, and teacher in Austin, Texas. She has a Ph.D. in theater history, literature, and criticism, with an emphasis in performance as public practice from the University of Texas at Austin.

1. "LGBTQ+ Definitions," Trans Students Educational Resources, accessed August 5, 2019, httsp://www.transstudent.org/definitions; "John Money, Ph.D.," Kinsey Institute, accessed August 5, 2019, https://www.kinseyinstitute.org/about/profiles/john-money.php.
2. As of November 2019, however, there is still no published list of all the options, making it difficult to find and change your gender on the platform. Leslie Walker, "How to Edit Gender Identity Status on Facebook," *Lifewire*, September 28, 2019, https://www.lifewire.com/edit-gender-identity-status-on-facebook-2654421.
3. Hope, "She Invented the Gender Reveal Party."
4. Landyn Pan and Anna Moore, "The Gender Unicorn," Trans Student Educational Resources, accessed October 2, 2019, http://www.transstudent.org/gender/.
5. The definition of "two-spirit people" is dynamic and not uniform across Native American peoples. There is also some controversy within Native American communities about two-spirit as an umbrella concept for many different gender identities. Unfortunately, the term is often appropriated by non-Native Americans to mean having both masculine and feminine traits, which divorces the term from its cultural specificity. Rebecca Nagle, "The Healing History of Two-Spirit, A Term That Gives LGBTQ Natives A Voice," The Huffington Post, accessed September 30, 2019, https://www.huffpost.com/entry/two-spirit-identity_n_5b37cfbce4b007aa2f809af1; Mary Anne Pember, "'Two Spirit' Tradition Far From Ubiquitous Among Tribes," Rewire News, accessed September. 30, 2019, https://rewire.news/article/2016/10/13/two-spirit-tradition-far-ubiquitous-among-tribes/.
6. Recent examples include the media scrutiny of the perceived masculinity and/or sexualization of black women athletes Caster Semenya and Serena Williams: the former went through several bouts of biologically invasive sex verification testing, and the latter's body and dress are under constant surveillance by tennis authorities, fans, and the general public. See Anna North, "'I am a Woman and I am Fast': What Caster Semenya's Story Says About Gender and Race in Sports," Vox, May 3, 2019, https://www.vox.com/identities/2019/5/3/18526723/caster-semenya-800-gender-race-intersex-athletes.

Dr. Denise F. Su
Paleoecologist, Cleveland Museum of Natural History

20

Just a Paleontologist

Denise F. Su

"It's my pleasure to introduce our speaker tonight. She is a female paleontologist...."

Imagine the introduction above where the paleontologist is a man rather than a woman and substitute "he" for "she" and "male" for "female." Did that sound weird? Was there an instinct to omit the adjective completely? If you answered "Yes" to both of those questions, you are not alone. I cannot count the number of times that I have been introduced as a female paleontologist, and every time, I cringe. Putting an adjective in front of me implies that I am not the norm, that somehow I do not belong. Paleontology—particularly field paleontology—is dominated by men, and this is very much reflected in how I am viewed as a scientist.

Fieldwork is not easy—there is no running water, electricity, or other creature comforts of the modern world. Often its remoteness translates into situations that have real life-and-death consequences. In other words, the field is a dangerous and difficult place to work, even for an experienced field paleontologist.

I know that I am not alone among my female colleagues when I say that there have been times when I pushed myself to, and sometimes beyond, the very edge of my endurance because I did not want to be seen as weak or undeserving of being in the field, or for my actions to somehow reflect poorly on all women who are in field paleontology.

Why impose such severe self-judgment? Because I, like many women who conduct fieldwork, have experienced firsthand the assumptions about women in the field: we are not strong enough, we require special accommodations, or we can't really be the ones in charge because it's fieldwork—a man's work.

My primary field research is in Tanzania, and this adds an additional layer

of complexity—specifically, our local collaborators' (local government and museum representatives and scientists) perceptions of my role. Because I am a Western woman, I am not treated the same way as a local woman. This, however, does not stop them from looking at me when anything that falls within the realm of "women's work" needs to be done. I generally decline on principle, as none of my male colleagues would be asked to do those tasks.

I am actually one of the lucky ones. I had a supportive Ph.D. advisor (male) and tremendous mentors (male and female) who did not have assumptions about women in the field and did not think that adjustments made for female scientists were special accommodations compared to those made for male scientists. And yet, I still felt the sting whenever someone turned to the male graduate student for instructions or I was dismissed as not knowing what I was talking about in spite of decades of experience. It would have been easy for these experiences to turn me off of fieldwork, but for the fact that being in the field and conducting my research was as essential to me

Dr. Denise Su takes the film crew behind the scenes at the Cleveland Museum of Natural History.

as breathing. Without that, I might have left field paleontology.

Because of my own experiences, I am focused on the next generation and helping the women who follow, sometimes by uncomfortably putting myself in public view just to say: "I went through what you are going through. I understand, and I am here." Documentary films like *The Bearded Lady Project* are invaluable. They not only make us ask hard questions of ourselves but also tell future scientists who might feel like they don't belong that there are others like them—and that they do belong.

And there is progress. When I first started doing fieldwork, I was often the only woman on the team. Now I am joined by colleagues or students who are women. This changes the dynamic in the field, and just as important, it normalizes the presence of women in paleontology and their experiences. This is the future—when the experiences of women are just as valid as those of men; when there is no judgment based on sex or gender; when no adjectives are necessary and I am just a paleontologist.

Denise F. Su, Ph.D., is curator of paleobotany and paleoecology at the Cleveland Museum of Natural History.

Dr. Carrie Tyler
Invertebrate paleontologist, Miami University

Dr. Kate Bulinski

Invertebrate paleoecologist, Bellarmine University

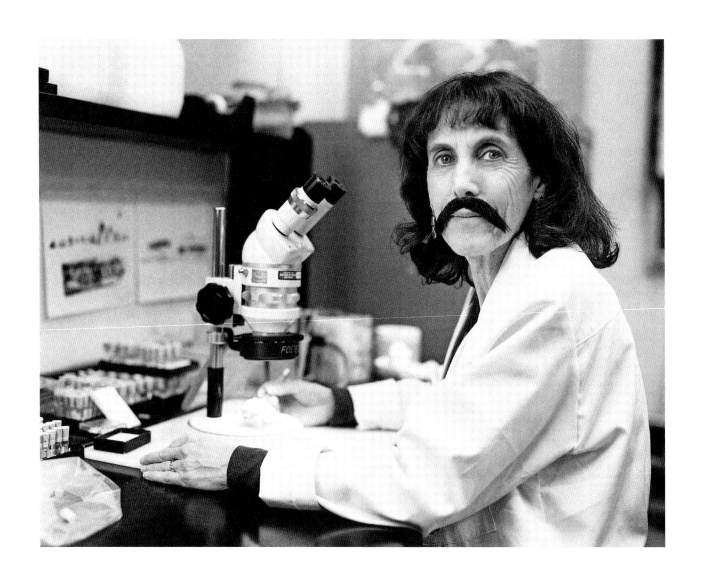

Dr. Alisa Winkler
Vertebrate paleontologist, Southern Methodist University

Dr. Kinuyo Kanamaru
Paleoclimatologist and paleotempestologist, University of Massachusetts

21

Performing Gender in Paleontology

Amy K. Guenther

Poised. Rugged. Serious. Intense. Smiling. Laughing. The faces looking back at you seek to upend the gendered stereotypes of paleontologists as venerable, bearded (white) men. The women pictured in *The Bearded Lady Project* portraits still wear the same clothes, body postures, demeanors, and hairstyles as they did before the camera crew walked into their research fields and labs. The only artifices actually visible are the carefully selected and trimmed beards, goatees, and mustaches applied moments before. Amused at being asked to perform gender in what, for most, is an unfamiliar, theatrical way, some smile, their faces stretched in new, exaggerated directions by the glue holding their artificial beards in place.

The Bearded Lady Project makes visible, through explicit gender play, the ways women must negotiate gender performativity in the sciences, specifically in the field of paleontology. Does a good paleontologist need a beard? No, of course not. But the placing of beards on women scientists is a cross-gender performance seeking to evoke a playful response to the real and sometimes very hazardous realities of performing gender in a field dominated by men.

What does it mean to perform gender? In 1988, philosopher Judith Butler began popularizing the concept of gender performativity. She explained that "gender is in no way a stable identity or locus of agency from which various acts proceed. Rather, it is an identity tenuously constituted in time—an identity instituted through *a stylized repetition of acts*." Through the repetition of acts, gender is not stable (for an individual or society), biological, or even binary, but multifaceted, malleable, and dialectic. Butler bridged a discussion of gender in phenomenology and feminist theory with the burgeoning field of performance studies. Performance scholar Diana Taylor defines the use of the word "performance" in performance studies as the event/object to be analyzed and the "methodological lens that enables scholars to analyze events *as* performance."[1] This latter definition better encapsulates the ways gender is performative.

Performance, like gender, is always in a process of becoming. Performance scholar Elin Diamond describes it as "a doing and a thing done," wherein "performance" "describes certain embodied acts"—the doing—as well as the "completed event...remembered, misremembered, interpreted, and...revisited across a pre-existing discursive field"—the thing done. Similarly, Butler asserts that "the body becomes its gender through a series of acts which are renewed, revised, and consolidated through time." Both definitions grapple with how to define something that is always in a process of becoming before, during, and after an individual enters the stage. In the performance of gender, the repetition of gestures, costumes, behaviors, etc. crafts a framework of being and moving through the world into which an individual enters and exerts influence.

Today, when scholars and activists argue that gender is performed, they do not mean that changing one's gender identity is as easy as changing one's clothing, a mischaracterization often promoted by gender essentialists—people who believe gender is entirely dependent on biological sex. This intentionally obtuse understanding of gender performance informs most of the transphobic bathroom legislation in various U.S. states whereby, they assert, a man in a dress will be free to assault women and children. As gender theorists like Butler and the lived experiences of feminists and LGBTQIA+ communities can attest, in any given hegemony gender is often not consciously or intentionally performed until society judges one to be performing it incorrectly. Thus, the use of "performance" in discussions of gender is not frivolous, insignificant, or capricious. Performing gender incorrectly in a given time and place has very real and sometimes very dangerous consequences.

That said, performances as framed events in time and space, like *The Bearded Lady Project*, often call attention to, subvert, accentuate, and/or poke fun at the gender performativity in a given time and place. Placing beards on women's faces plays at cross-gendered casting, wherein the actor's gender does not match the character's. Here, the women scientists are cast as the stereotype of a scientist, a man whose beard signifies his maleness. Modern feminist and/or queer theatrical productions often use cross-gender casting to accentuate the character's gender role in the world of the play; to call attention to the perceptions of gender that the audience brings into the theater; and to reflect on the contemporary moment's social construction of gender. This is explicitly different from when women were legally barred from performing on stage and men performed all the roles during various times in Western history. Seeing a woman actor perform a male role is an example of what playwright and theorist Bertolt Brecht calls "alienation," whereby the audience is forced into critical aesthetic distance from the believability of the world of the play and does not fall into the passive trance evoked by realism.

The participants in *The Bearded Lady Project* photographic exhibition are not attempting to pass as men. They simply wear the clothes they were already wearing in the field or lab. The clothing choices for the outdoor field are largely gender neutral or leaning toward masculine out of practicality. They are meant to be breathable and easy to move in, so they are not tailored close to the body. Pants, long sleeves, and hats protect the body from sunburn, bugs, and the scrapes of rocks and brambles.

For many of the subjects, the choice of outdoor clothing reflects an additional safety concern. Being perhaps the only woman camped at a particular site for days or weeks carries similar safety warnings as women receive when walking alone at night: do not draw attention to yourself. The clothing of the field often acts like camouflage, deflecting attention away from their status as women, alone.

The beards, carefully selected and attached off camera, blend into the overall aesthetic of the black-and-white photographs they parody. If you glance at the portraits quickly, as I have seen several observers do, you might assume they are of the famous male paleontologists represented in textbooks. But if you look carefully, you see that these paleontologists are women, not men—not women trying to pass as men. You can see the artifice of the beards, the netting attaching the hair to the face, mimicking the subtle exposure of artifice that is at the heart of the entire project. The attachment of a beard to these field costumes both critiques the expectation that paleontologists are men and makes explicit in an exaggerated fashion the implicit gender bias in the sciences and the more subtle clothing and behavior choices women paleontologists must make to survive conferences, classrooms, and academic departments.

The stories told in *The Bearded Lady Project* demonstrate that many of its subjects were forced into a more conscious gender performance than was necessary before they became paleontologists. Women paleontologists face many of the same biases as women in other careers, but with a special emphasis on more masculine clothing and traits. Well-meaning but misguided mentors still advise women paleontologists to downplay any performances of femininity: tight-fitting clothes, skirts and dresses, bright lipstick, substantial makeup. This recommendation has several problems. It presumes that women are inherently drawn to femininity. It is also given in the guise of professionalism, wherein "professional" is a code word for more masculine. Like the clothing of the field, it implies safety concerns, i.e.,

Practicing the beard
stroke in Lyme Regis

that women who dress a certain way are asking for harassment. Anecdotes abound of early career women paleontologists forgoing certain conferences and even changing their specializations to protect themselves from the advances of older, gatekeeping men. The implicit and sometimes explicit messages women paleontologists receive are to be more masculine and less explicitly feminine to earn more respect and, more insidiously, to protect themselves from sexual advances. This is not to say that all women paleontologists gravitate toward performing femininity, but that all paleontologists enter into a system that implicitly judges them for their adherence to a certain gender performance. Performing well in the sciences should not be tied to the perception of performing gender correctly. Yet the implicit message is to exaggerate the masculine over the feminine.

Today, however, many women paleontologists are consciously performing high femme at conferences and in the classroom because they like makeup and fashion, find it empowering, and do not see this as mutually exclusive with being a well-respected paleontologist. Several *Bearded Lady Project* subjects have described these performances as conscious decisions to emphasize their femininity, and their status as women reflects larger movements in feminist and queer discourses that advocate for wearing what makes the wearer feel good regardless of hegemonic gender roles. That is, expressing femininity through clothing does not mean taking on the more negative stereotypes of femininity—as passive, weak, or less intelligent. Moreover, many of the portraits taken in labs rather than in the field feature subjects with more feminine-leaning signifiers in their clothing and hairstyles. This is not to say that good paleontologists should be more feminine. Instead, it complicates the notion that there is any one good way to perform paleontologist.

Paleontologists are adept at reconstructing ancient worlds from the fossilized bodies of plants and animals, but many are still learning to engage with how their own bodies create meaning within their disciplines. *The Bearded Lady Project* photographic exhibition includes a variety of bearded ladies, not just those parodying the nineteenth- and twentieth-century photographs of ye olde (bearded, male) paleontologists, and in doing so expands the definition of what performing gender in paleontology could mean.

Amy K. Guenther (she/her) is a freelance scholar, dramaturg, and teacher in Austin, Texas. She has a Ph.D. in theater history, literature, and criticism with an emphasis in performance as public practice from the University of Texas at Austin.

---

1. Diana Taylor, *The Archive and the Repertoire: Performing Cultural Memory in the Americas* (Durham, NC: Duke University Press, 2003), 3.

Sophie Westacott

Ph.D. candidate in paleontology, Yale University

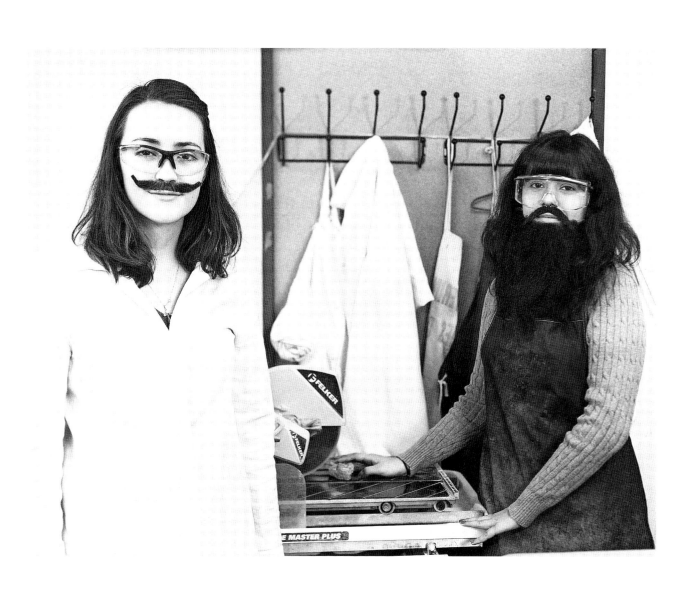

Elizabeth Stephani and Sabrina Cordero

Biology undergraduate students, Smith College

Linnae Rondeau

Ph.D. candidate in geoscience, University of Massachusetts

Dr. Margery Coombs
Emeritus professor of biology, University of Massachusetts

Adriane R. Lam, Raquel Bryant, and Serena Dameron

Leckie Lab, University of Massachusetts Amherst

Dr. Diane Erwin

Senior museum scientist, University of California Museum of Paleontology

Women in paleontology at the University of California, Berkeley

Dr. Sara B. Pruss

Paleontologist, Smith College

22

Taking Off the Beard for Good

Sara B. Pruss

As you will notice, my portrait is different from the others: I am not wearing a beard. How does a beardless woman end up participating in *The Bearded Lady Project*? My explanation is this:

I was inspired to first learn that this project would highlight the struggles women have faced in science, particularly in geosciences, which is a traditionally male-dominated field. The project's aim was to provide voices to women across generation and rank, and this resonated with me strongly. Women who had been graduate students twenty years ago and women who were graduate students now finally had a chance to tell their story. And so in late December 2015, Lexi Marsh and her team arrived at Smith College in western Massachusetts to interview and photograph me with my then-current students and a lab alum who was visiting (Sophie Westacott, Yale University). I was delighted that someone took the time to include me in this project, as a woman and a paleontologist teaching at a women's college. During the interview, I explained that my portrait would be different from the others: I would not be wearing the beard.

My decision not to wear the beard was complex and arose from feelings that I could barely articulate initially. Regardless, Lexi and her team supported it, but they did want me to explain why I had decided not to be photographed wearing a beard in the caption that would appear with my portrait. They also asked that I further expound on the topic in this essay.

Teaching at Smith College has forever changed me. We often reflect with our graduating seniors on how Smith has changed *them*, as they move on to face a world that is quite different from the college. But I am not the same person who arrived on campus in 2007. Before then, I had never taught a transgender student. Nor had I ever been in the field with a group that included nonbinary students, or heard their stories, or understood their journeys. As I was faced with putting on the beard, I felt again as if I were

being asked to support the notion of a world in which there are only men and women scientists, and I couldn't help feeling like there were entire groups of my students being left out of this conversation.

While I understood that the intent of putting on the beard was to shed light on how ridiculous it is that women—or anyone, really—should have to modify themselves in some way to feel accepted in their field of study, this felt like a perpetuation of a norm that I have come to reject. So in my portrait I appear just as I am. And my hope is that my students who go on in the geosciences can always be just as they are and still be greeted with the acceptance that we all hope to feel in our careers.

I count myself among the lucky women in geosciences who have had a productive career, buoyed by the support of my male and female colleagues, my students, and my institution. Teaching at a women's college, where I have had the pleasure of being an unencumbered feminist scientist for my entire career, has made me acutely aware of my fortunate circumstances and the need to give back to my community. Many members of groups who are underrepresented in geosciences do not receive the support that I had: colleagues who took my students into the field or collected samples for me when I was home nursing a newborn; made sure I received prominent

Dr. Sara Pruss sits for her interview in her lab at Smith College.

roles at conferences, an opportunity that the sometimes invisible liberal arts college professor misses; and nominated me for awards in my field. Even so, I, like every woman scientist I know, have felt overlooked, pushed aside, and ostracized by male colleagues. For these reasons, my hope is to be a role model, supporter, and ally to all marginalized groups in our field, especially now as a senior scientist. We have much work to do as a community, but I think we are moving toward a place where we can take off the beards for good.

Sara B. Pruss, Ph.D., is a professor of geosciences at Smith College. She earned her undergraduate degree in biology-geology at the University of Rochester and her graduate degrees in geological sciences at the University of Southern California, and she was an Agouron Postdoctoral Fellow at Harvard University prior to arriving at Smith College.

PART III
Behind the Lens

*It takes a filmmaker to fight for a paleontologist.
It takes a paleontologist to risk everything
in support of a filmmaker. It was friendship that
founded this project.*

Lexi Jamieson Marsh

Lexi Jamieson Marsh
Director and producer of *The Bearded Lady Project*

23

Behind the Lens: Filming a Documentary

Lexi Jamieson Marsh

The role of a documentary filmmaker is to capture life's authentic moments in a visual landscape, and from it piece together a cohesive and compelling narrative.

The filmmaker's purpose is to transform a microcosm of reality into a larger conversation, in the hope that viewers will gain new insights and find personal connections.

The filming of *The Bearded Lady Project* documentary took over two years and resulted in eighty-three hours of video footage. It took three months to edit the documentary short film. Two years later, the feature-length film was finished. And yet, after all that time, there were still more stories to tell.

What I have found to be unique to documentary filmmaking is this: The camera records a real-world image with an unchanging time stamp. The completed work remains forever set in a particular time and place. The storyline, composed of raw video files, is shared in a format that reveals contemporary truths. Selecting what is revealed and determining what remains untold are equally critical to the creative process. As the director, I discovered that as I made a deeper connection with the subjects, I needed to balance the dual requirements of honoring the community of scientists and telling a compelling story. This is the challenge that faces every creative artist, certainly, but it is even more challenging for the documentary filmmaker. We seek to present reality while also discerning the hidden truth beyond the visual image.

The Bearded Lady Project was the ideal opportunity to work on a topic I was passionate about, with people I greatly admired. I started with a micro crew made up of friends and sponsored by family. In the beginning, I was uncertain how far I could go. But the journey and the final products have

Lexi Jamieson Marsh and Eliana Álvarez Martínez prepare for an interview while filming on location in Raymond, Washington.

greatly exceeded all of my expectations. The expansion of the project could only have been achieved by pushing far beyond our time and financial limits; this was in part accomplished with the support of fellow artists and scientists. The reception that *The Bearded Lady Project* has received has promoted a genuine dialogue about the role of the scientist—that has been the most rewarding aspect of this project.

Looking back to the start of *The Bearded Lady Project*, when we officially moved forward with our first shoot, the visuals I had in mind were clear: I wanted to highlight women at work. This story would be shown by recording the scientists physically interacting with the landscape. I wanted to document what fieldwork entailed, to see a day in the life of strong women who were confident in performing their daily tasks. In Hanna Basin, Wyoming, where our very first shoot took place, my "to-do" list included a variety of close-ups: shots of hands digging, hammering, and brushing. For example, we captured heavy-booted feet clambering up mountains and descending ravines in active pursuit of fossils. I wanted to start the film by connecting the human body to the landscape, providing texture and contrast to the scientists' practical efforts. The women who were filmed were not separate from their work; they were physically attached to their research.

150

Dr. Kay Behrensmeyer unveils a sample for Cinematographer Draper White.

Another aspect of filming was the interview. I began each one by asking the scientist about her research. This was an ideal icebreaker; it allowed the subject to relax into the process of being interviewed. While working with paleontologists in the field, I was surprised by how easy it was to film them. They were ready to roll whenever I was and were undaunted by changes and adjustments. However, as filming continued, my original outline no longer worked. The creative palette grew to include many more participants than I had originally envisioned. Holding fast to my original story would never have allowed the flexibility to include this number of emerging stories.

It was during our shoot at the University of California, Berkeley, where I had planned for two interviews and three portraits, that the project truly expanded. At that point, in February 2015, we documented more than thirty female paleontologists during our three-day shoot. Instead of limiting my focus, I decided to take the opportunities that were coming my way to expand the story. The Berkeley shoot shaped my concept into a much broader view. It was a pivotal moment in this project, and it increased my awareness of the number of scientists acknowledging these issues. Sharing their varied experiences became the compelling throughline of *The Bearded Lady Project*.

As the story took shape, I gained increasing confidence. After almost a full year, I was finally able to sit back and allow the story to be told to me, instead of trying to take control. I had set out to document the wide variety of females working in paleontology, to challenge the assumption that only women who identify as "tomboys" would want to pursue a career working in the wilderness. In addition, and somewhat to my surprise, I discovered that women paleontologists share core similarities with film directors. Moreover, the more we looked at the experience of women working in paleontology, the more we realized that this particular field of science could be used to examine the struggles of women in all sciences, including technology, engineering, and mathematics. Documentary film opened up the floodgates to examine all other stereotypically male-dominated fields into which women had strategically inserted themselves. If we could succeed by standing up to the traditional image of a paleontologist, what other negative images could we confront?

The conversation was now expanding well beyond the world of paleontology. By bringing this particular issue of pervasive negative stereotypes in paleontology into a wider public discourse and doing it in a creative way that invited reactionary commentary, we were able to not only question the societal representation of women but also reflect upon how we represent ourselves in society.

As quickly as the project grew, word got out. Before it was completed, our team was presenting at national meetings and conferences. We were given more names of scientists who should be interviewed, whether they wanted to participate or not. Suddenly, while entrenched in the creative process, we were becoming aware of the attention we were garnering—and with attention comes opinions. As with any creative endeavor, there is an inevitable moment of doubt. With documentary film, the ethical and moral dilemmas should always exist. But, we wondered, would those questions overwhelm the film and even cast doubt on the purpose of its creation? The simple act of being photographed wearing a beard—a comical and unprofessional endeavor—can potentially dismantle decades of groundwork laid to establish one's competency in the workforce. Women need to be taken seriously; professionalism in the workplace means conforming. To step out of line, to show that you are different, risks ostracization. Unbeknownst to me, this is what Ellen Currano was facing a year after we began shooting.

In September 2015, our film crew returned to Wyoming to film Ellen's feature interview. What I hadn't taken into account was Ellen's hyper-self-awareness, which had grown since our first shoot. She had come to the realization that she was not just a participant—she was the face of the entire project. Her involvement meant that her world merged with the project,

Silly moments during the very first bearded photo shoot in the Hanna Basin, Wyoming

and when it came to her career, the two would be impossible to separate. I found myself completely in the dark in trying to understand Ellen's experience. My own experience of making this film was completely different. Of course, I had a lot less at stake.

The truth hit me hard. I was trying to get a good interview from my friend who knew that her words would linger long after our interview ended and the cameras stopped recording. She knew her colleagues would be watching, and that what she said would follow her and potentially discredit her research. If I made a film that was not well received, I could still go about my life relatively unscathed. Ellen, however, had real personal risks: it was her face on the poster; it was her story that was going to be told and retold. Ellen's internal concerns about the reception of the final film were confirmed when a colleague recommended that she start to focus on something other than *The Bearded Lady Project*. He emphasized that he supported her efforts and believed them to be needed; however, the work Ellen had put into outreach would not be considered for tenure. The project was taking time away from conducting science.

When making art, the first part is figuring out how to do it. Yet, as Ellen's example demonstrated, I was now struggling with the implications of what would happen if I finished it.

Over the course of filming, the beard became a visual metaphor for the obscure and unfounded reasons for keeping women away from science: the lack of facial hair was the only feature that made the women's existence a liability in the field. As silly as putting on a fake beard is, the truth to the story—to the history of women in science—is that they have had to alter who they are, to put on a mask in order to be accepted as equal in a male-dominated field, to hide that they are outsiders. Through the film, we see these women conducting science, working in an environment that is both inspiring and rewarding. Yet over the course of the film, we come to understand how they got to the point of applying the beard. The participants collectively participated in a physical act of defiance regardless of expertise, geographic location, age, or race. How did that act change their understanding of the world and how they exist within it?

We have had many varying responses to *The Bearded Lady Project*, both positive and negative; that is the purpose and the challenge of making art. This is also the wonder and beauty of it—there is no singular response, no "correct" way to process this information. Regardless, we need emotional investment of some kind if we want to right wrongs. Opinions matter. The power of documentary is to make individual lives universal—to show the world through another person's eyes and challenge the accepted reality, one bearded lady at a time.

Lexi Jamieson Marsh is the founder of the independent production company On Your Feet Entertainment and the director of *The Bearded Lady Project* short and feature-length documentary films. She is currently visiting assistant professor of media and culture at Miami University.

24

Being with Artists in the Field

Ellen Currano

Artists and scientists are stereotypically viewed as polar opposites, exemplified by the old notion that people are left-brained (scientists, engineers, mathematicians) or right-brained (artists, writers, musicians).

In reality, scientists and artists have overlapping skill sets, and there is incredible potential for collaboration—so much so that STEAM, short for science, technology, engineering, art, and mathematics, is a thing. The integration of art and design into STEM research and education is widely supported, from the U.S. National Science Foundation to the U.S. House of Representatives to *Sesame Street*.

Both science and art require creativity, innovation, keen powers of observation, and curiosity about how the world works. This was all clearly on display when *The Bearded Lady Project* film and photography team joined me in the field. Once I got over having a camera in my face all the time and having to walk the same path, swing the pickaxe yet again, or repeat the same motions so the film crew could get the shot from a different angle, I found myself really enjoying having fresh sets of eyes out in the field with me. In particular, Cinematographer Draper White peppered me with questions about the landscape we were walking across—the ancient landscape we were reconstructing using the rocks and fossils and the methods we used to do the reconstruction. Draper also proved to be an adept fossil finder: we had to literally pull him off an outcrop where he was having great fun—and success—looking for tiny fossilized shark's teeth.

In retrospect, both Draper and I benefited from our conversations. He got a free crash course on the natural history of the Rocky Mountain region, while I found that, in order to answer Draper's questions, I had to think

Draper White

Cinematographer and portrait photographer, *The Bearded Lady Project*

more deeply about my science, put my answers into a layperson's terms, and think of modern analogies that would allow me to verbally paint a detailed picture of what had occurred beneath our feet over fifty million years ago.

Paleontologists collect and analyze fossils. Artists create imagery. By collaborating, they can bring extinct and exotic creatures, catastrophic asteroid impacts and volcanic eruptions, ice ages and greenhouse intervals, and so many other aspects of Earth's history to life and capture the imagination of the young and the young at heart.

Ellen Currano, Ph.D., is a paleontologist whose painting of a tent under the stars was mistaken by her partner for a tongue under the stars. Although she is confident that she could become a better artist with practice, she instead devotes her time to science, outdoor adventures, and collaborations with professional artists.

Kelsey Vance
Portrait photographer, *The Bearded Lady Project*

25

Creating Portraits for The Bearded Lady Project

Kelsey Vance

At the outset of *The Bearded Lady Project,* my idea for photographs of paleontologists was rooted in the past: the iconic black-and-white portrait of a man or a group of men in a rocky landscape with several days of dirt embedded in their clothes, holding tools for digging.

With that in mind, I had the inspiration to change the image. I knew without question that I would create the portraits with the same type of camera used over a hundred years ago to document the uncharted territories of the western United States (work that helped establish our National Park System): a large-format field camera complete with a bellows, a dark cloth under which I could examine the image on the ground glass, set atop a sturdy tripod.

The large-format camera is undoubtedly a cumbersome system for trekking through the field, cramming into a small lab, and transporting through modern airport security systems (the camera definitely looks suspicious when viewed by an x-ray machine, and the boxes of film had to be hand inspected but couldn't be opened, as that would ruin the film and latent images). However, the large-format camera creates a unique experience for the subject. It is a beautiful object and always causes excitement and curiosity from the people being photographed. Furthermore, it is my favorite camera to use for its elegance, technical precision, and incredible image quality.

The *Bearded Lady* portrait series represents a collaboration between subject and photographer. Where people stood, what they wore, the tools they carried—these were all of their own choosing. The only alteration, not required, was the addition of facial hair to reference the predominance

of men in the field of paleontology. The large-format camera also encouraged teamwork and partnership, as the subjects remained relaxed while I composed the image under the dark cloth. Once the frame was set, I stayed engaged with the subjects by standing beside the camera rather than pulling it up to my face to capture the picture. The entire process is slower, more informal, and more mindful than with a handheld digital camera.

Dr. Catherine Badgley stands for her bearded portrait in the Mojave Desert, California.

Technical Information

The camera I used to create portraits of the women featured in *The Bearded Lady Project* is a Linhof Technika IV, dating back to the mid to late 1950s and built to last for the ages. Most cameras being used today are digital, but film still exists. This particular camera uses 4×5-inch sheet film. Each sheet of film is placed into a cartridge that is loaded into the back of the camera one frame at a time. I typically made four images (four sheets of film) per subject. The advantage is a large negative, enabling one to print portraits very large while maintaining good image quality. For this project, I used Kodak 100 and 400 TMAX black-and-white sheet film.

After the film is exposed to create an image, it is kept in complete darkness until fully developed. Each of the negatives was hand developed, fixed, washed, and air dried in a traditional analog darkroom in the basement of my home. I then placed the negatives in direct contact with photographic paper to produce contact sheets, which allowed me to see the images in their positive form and begin the editing process.

This is where the process entered the twenty-first century: the selected negatives were sent to a lab for digitizing through high-resolution scanning by a Hasselblad Flextight X5 scanner, a machine made specifically for scanning film. After the images were digitized, I viewed them in Photoshop—as one would in a darkroom—eliminating any dust, correcting the exposure, etc. Once this last review of the photographic portraits was complete, they were ready to be printed for the catalogue and for exhibition display.

Gear Highlights
- Camera: Linhof Technika IV
- Film: black-and-white 4×5-inch sheet film (predominantly Kodak 100 and 400 TMAX)
- Film scanner: Hasselblad Flextight X5

Exhibition Prints
- Canon image PROGRAF iPF8400 printer with LUCIA Pigment Inks
- MOAB Entrada Rag Natural Bright 290 gsm paper
- Each print measures 24×30 inches and is in an edition of five

Kelsey Vance is the photographer for *The Bearded Lady Project*.

FIELD NOTES

Lexi Jamieson Marsh

2014

JUL 22

Hanna Basin, WY: Marieke Dechesne was one of the first scientists to join our project. She was asked to do so while working in the field. Concerned about the response from her institution, Marieke asked if she could be photographed without the beard. This is a stunning photograph that demonstrates the freedom offered to women in the field—a windblown scientist who is nevertheless very content and has no place she would rather be.

JUL 24

Hanna Basin, WY: Timing is everything. Dr. Penny Higgins joined *The Bearded Lady Project* on this day, our very first shoot. Up to that point, we did not have a website or a trailer—and we didn't even give her much of an explanation of the project, other than an email to request her participation. To our shock and joy, Penny came on board. When Draper White, *The Bearded Lady Project*'s cinematographer, drove into camp the first night and saw Penny on top of a hill wielding her very own broadsword, we truly knew she was the perfect scientist to help us challenge gender stereotypes in paleontology.

JUL 24

Hanna Basin, WY: Routine is part of working in the field. For Jenna Kaempfer, braiding her long hair was part of her morning ritual. In the field, bathing is not going to happen regularly, so keeping long hair knotted together will ward against unruly tangles (as our crew personally experienced). Jenna wasn't interested in participating at first, but when her professor, Dr. Penny Higgins, stepped up to the plate and donned a long bushy beard (and owned it), Jenna asked if she could wear one too. She only had one condition: could she braid her beard?

2015

FEB 8

University of California, Berkeley: Dr. Cindy Looy was the first scientist to make a personal beard request. As a *Big Lebowski* fan (she even named a scientific fossil discovery *Lebowskia grandifolia*), Cindy requested a beard that resembled The Dude's. And we did our best to abide, finding a beard that looked just like his.

FEB 10   **University of California Museum of Paleontology, Berkeley**: Dr. Carole Hickman agreed to sit for a portrait under one condition: she would supply her own mustache—the same mustache she'd worn while working in the Australian outback in the 1970s. Back then, she wore fake facial hair to ward off strangers who would interrupt her fieldwork because they assumed she was a helpless woman stranded in the wilderness. Once the mustache went on, she was able to conduct her research in peace.

FEB 10   **University of California Museum of Paleontology, Berkeley**: Dr. Lisa White is the assistant director of education and public outreach for the University of California Museum of Paleontology. It is a research museum where a lot of the specimens remain out of sight to the general public but are available to students and faculty for their own research. Surprisingly, Lisa says, a lot of these fossils are understudied. "All of these specimens deserve a second look [for] new approaches to interpreting them," Lisa noted. Her portrait was photographed just outside the museum's entrance; you can see the foot of the magnificent *Tyrannosaurus rex* welcoming visitors.

FEB 11   **University of California, Berkeley**: Leslea Hlusko's lab was one of the most aesthetically pleasing interior locations we filmed during our shoot. It featured uncharacteristically large windows for a lab; this brightened the space, and the light streaming through floor-to-ceiling glass display cases illuminated her collection of primate skulls. When we selected the location for Leslea's portrait, it was obvious we needed to have her and her graduate students pose in front of the skulls.

APR 20   **Rainbow Basin, Mojave Desert, San Bernardino, CA**: Shooting a documentary about paleontologists means filming in remote locations around the world. In the case of Dr. Catherine Badgley, our film crew traveled to the Mojave Desert. We set up Catherine's interview in a picturesque location near the Rainbow Basin, a stunning rock formation with a blend of colors that resembles a rainbow. In order to capture the beautiful landscape, Catherine agreed to perch precariously on a boulder, but she made balancing on it look easy—maybe even comfortable. In the middle of our interview, there was a loud bang, as if a thunderstorm were crashing overhead. Our team looked around, but

there were no clouds to indicate where thunder could have come from. "Oh, that was just a sonic boom," Catherine calmly informed us, clearly not concerned by the sound. "There's a military base not too far away."

APR 20 **Rainbow Basin, Mojave Desert, San Bernardino, CA**: Matching beards to hair was more tricky than we originally expected—especially when scientists didn't send us the most current picture of themselves. Dr. Katharine Loughney sent her photo with the caption: "[My hair is] actually pink right now." So our team contemplated looking for a pink beard, or even dyeing one of the beards we had. Although the idea was fun, we decided that with the black-and-white film we were using, the result would not be as impressive as we would have hoped.

APR 20 **Rainbow Basin, Mojave Desert, San Bernardino, CA**: When setting up for the bearded portraits, we encouraged the paleontologists to think of what they would like to feature in the frame; usually something important to them such as a tool or a personal memento. For Dr. Tara Smiley, it was her hammer, which she named "Lady."

APR 23 **Bisiti/De-Na-Zin Wilderness, San Juan County, NM**: Dr. Lisa Boucher met us at her field site in northern New Mexico a few days before her team arrived for their scheduled fieldwork. Her early arrival meant that after our film crew left, she would be camping alone for a few days while waiting for the rest of her team. This didn't seem to faze her at all. During her interview, she told us that women shouldn't be afraid to camp alone; it all comes down to being prepared and aware of your surroundings. Time was limited, so Lisa's interview was scheduled for sunrise the next morning. Lisa was all about tradition, waking up a little earlier in order to have the one luxury she allows herself while working in the field: powdered chai latte.

MAY 5 **Raymond, WA**: No two field sites are the same. Although this was only our fourth shoot for what would become a two-year production, the requirements were much different from the rest, which were mostly located in dry deserts. Here, in a Pacific Northwest salt marsh, we needed waders. And we needed to be mindful of where we walked. Andrea made sure our crew was well aware of the hazards before we ventured into the field with her: "I've known people to break

their legs in a salt marsh." Not only was the mud treacherous for walking, but it also suctioned around our feet as we stood. Cinematographer Eliana Alvarez Martinez was filming with us and had been focusing on the work being conducted by Andrea and her team. After getting the shot, Eliana instinctively took a step forward, only to realize, as she fell, that the mud held her feet in place. She reacted by lifting her hands high above her head, crashing face-first into the marsh so that her camera could be saved from the mud, which would have destroyed it.

JUL
7 **Exeter University Devon, southwest England**: Dr. Claire Belcher had recently purchased a Pomeranian puppy named Alby before we arrived in England. She was kind enough to drive us from Exeter to film in Lyme Regis for the day, and since we wouldn't be back until late, the puppy came with us. Claire, like most paleontologists, came prepared, with all the supplies needed for him to stay comfortable during his day in the field. After reviewing the photographs, we decided our final shot should include Alby. When it came time to select *the* portrait we would use for the exhibition, the one with Alby was undeniably the best.

JUL
8 **Lyme Regis, West Dorset, England**: Our film crew met Dr. Kate Acheson outside the Lyme Regis Museum in the south of England on a very bright and hot day— so bright that the film director got one of the worst sunburns of her life. Kate was in the area working as a local tour guide, leading families down the rocky beach to help them search for fossils. We had filmed and photographed our subjects in remote locations and laboratories; this was the most populated location to date. When it came to wearing her beard in public, Kate was fearless, walking past scores of tourists enjoying their oceanfront holiday, now featuring a film crew and a bearded lady. Our walk was a long one; we carried our gear the length of the beach three times during the shoot. Since portraits can take a while to set up, Kate didn't waste a minute, hammering rocks in search of fossils. The beard barely fazed her; she kept it on long after we were finished taking her portrait.

JUL 8   **Exeter University, Devon, England**: Dr. Victoria Hudspith was very shy at the beginning of our shoot at the University of Exeter in England. Up to that point, all the participants had welcomed the idea of wearing a beard. Victoria, however, gave us concern that she might be feeling pressured to participate. Our crew always made sure to check in with the subjects to confirm that they were comfortable participating in this project. Victoria decided to go through with her portrait and in the end didn't want to take her beard off. During her individual interview for the documentary film, she shared with us that the project gave her some much-needed inspiration and perspective. She also gave us one of the most profound quotes of the film: she reminded us, and the audience, that we should be "telling those people who inspire us that they do inspire us, because maybe they don't even realize it."

JUL 8   **Lyme Regis, West Dorset, England**: What is important to a paleontologist? When it came to packing for fieldwork, we saw a wide range, from tools to personal objects. Each scientist selected a meaningful item to include in her portrait. For Dr. Jessica Lawrence Wujek, plush Elvis Bear needed to be featured. Jessica got Elvis Bear before her first trip as an instructor on a program called GeoJourney, where she led students in the field for nine weeks at a time. "It was one of the hardest jobs I had, but also one of the most amazing." Elvis Bear is well traveled: he's accompanied her to England, throughout the United States, and to Norway, Romania, and Germany. And he sings!

JUL 9   **Lyme Regis, West Dorset, England**: Sarah Baker, graduate student working in the wildFIRE Lab at the University of Exeter, was photographed on the coast of Lyme Regis, England—the field site of Mary Anning, the mother of field paleontology.

SEP 24   **Falls of the Ohio State Park, Clarksville, IN**: Dr. Kate Bulinski invited us to join her field trip. When we sat down with Kate for her interview, she discussed a moment when she enjoyed discovering the world: hiking as a kid, she found a black widow spider. "I ran to grab a stick so I could pick it up for a closer look." Her fascination with a spider was definitely not the answer we had expected.

OCT 18   **Southern Methodist University, University Park, TX**: "All the [treacherous events] you survive turn into great field stories." Dr. Bonnie Jacobs has over thirty years of African fieldwork under her belt and has accumulated her fair share of stories. The first near-death encounter Bonnie ever had was with a group of hippos. Arriving at camp in Kenya late at night, she had unknowingly pitched her tent between the hippos and their favorite food supply. "They're very quiet on their feet, so it's really like this pausing, ripping, gnashing sound. And all I had in my tent with me was a flashlight, so I didn't know what to do other than just sit there and be scared." Bonnie reminds us that while many would assume these lumbering creatures are harmless, "hippos are notorious for being dangerous to people; they're more dangerous than some of the big carnivores in Africa. So the hippo keeps looming and looming, closer and closer, until it gets right up to my tent. And right at the last second, it veers off and trips over the rainfly rope. The tent comes bulging in at me. It was pretty scary, so that was a good field story."

2016

FEB 12   **Wrightsville Beach, NC**: Flooding and high winds are nothing new to coastal North Carolina. We had originally planned to document Dr. Patricia Kelley on the beach; however, given the weather, we began to reconsider. Paleontologists are tough, but when Patricia wanted to move forward with the oceanfront portrait session in spite of the wind, we realized just how tough they are! Our team agreed, planning to head out earlier to see if we could beat or at least wait out the storm. After locating Patricia, we huddled in a car, dried the rain from her face, and applied the beard. We were lucky that the clouds parted long enough for us to set up the shot. We lost one umbrella and ended up using our bodies to shield the camera lens from the ocean spray, but in the end our efforts were successful. Patricia ripped her beard off, jumped into her car, and headed straight to her student's Ph.D. defense.

FEB 14 **Smithsonian National Museum of Natural History, Washington, DC**: Dr. Anna "Kay" Behrensmeyer decided to stand in front of a diorama for her bearded, or in her case, mustached portrait, which was shot in the storage space of the museum. The diorama she chose is a decommissioned reconstruction that no longer has a place in the public exhibitions but, thanks to Kay's husband, was removed from its storage crate for her portrait. The diorama holds a special place in Kay's heart: "I find [dioramas] very magical, I always have. They take you back in time and allow you to imagine being there." The two changes she would make for a more authentic landscape: "More broken branches and poop."

FEB 27 **Cleveland Museum of Paleontology, Cleveland, OH**: Dr. Denise Su is the curator of paleobotany and paleoecology at the museum. While filming Denise's interview, we asked her to describe what exactly had inspired her to become a paleontologist. Denise's eyes lit up as she went into detail about the collection of *Time Life* books her parents had when she was growing up. "My favorite was *Early Man* because it had this skull on the cover." It was the pictures that first piqued her curiosity about paleontology—that, and the fact that Denise's sister thought the book was too scary and would hide it from her.

OCT 18 **University of Texas at Austin**: Amy Atwater (pictured, right) is an advocate for mental health and awareness. She hadn't prioritized her mental health until she was doing her master's degree. Feeling like she was drowning, Amy sought out the support she needed in the form of therapy, self-care, and medication. From this process, she became aware of how she had always been a skilled and passionate paleontologist. Amy's efforts through social media and her blog, *Mary Anning's Revenge*, are bringing much-needed attention to the importance of good mental health.

DEC 22 **Smith College, Northampton, MA**: Dr. Sara Pruss is a faculty member at Smith College, an all-women's school. While she strongly supported *The Bearded Lady Project* as a way for women to tell their stories, she wanted to make a statement of her own by posing without a beard. She wanted to be shown exactly as she is, in support of all her students and their array of gender identities. She hopes that all students may feel inspired to pursue science exactly as they are.

DEC 22 **University of Massachusetts Amherst**: We had a last-minute addition to the project thanks to a recommendation from a helpful advisor. Our shoot was scheduled for 4:30 p.m., not ideal for parents with young children. Still, graduate student Kinuyo Kanamaru's children (both under six) kept themselves entertained with sample buckets and other tools around the lab. We are so pleased that Kinuyo was able to participate, and the shoot was greatly improved by the giggles of the children as they watched their mom put on a mustache.

2017

JAN 22 **Cleveland Museum of Natural History, Cleveland, OH**: Ashley Hall never sat for a bearded portrait. However, because she was supportive of *The Bearded Lady Project* early on and was tireless in developing outreach and public engagement during our second stop on the official exhibition tour, she is an honorary Bearded Lady. Ashley is an incredible advocate for science. Through social media, she became an early adopter of our project and really helped us establish a creative conversation within a scientific setting. Ashley engages with the public to make science accessible through social media. Follow her on Twitter: @LadyNaturalist.

MAR 7 **APR 16** **Golden, CO,** and **Roseburg, IN**: Although many of the field sites our film crew visited during the production of *The Bearded Lady Project* were remote, Dr. Karen Chin's and Dr. Carrie Tyler's (pictured right) sites (Colorado and Ohio respectively) were very accessible. In fact, had we turned the camera around at either site, you would have seen a paved road. At both sites, the construction of highways, which were carved out of rock, exposed new outcrops of fossils: dinosaur footprints at Karen's site and trilobites at Carrie's. You too may have fossil outcrops along your daily commute.

WHY DIVERSIFY SCIENCE

Unconscious biases, institutional practices, and even the way academia and the broader scientific community are structured make it more difficult for women, people of color, the LGBTQ+ community, people with disabilities, and other underrepresented groups to become scientists. Many of the scientists who agreed to become Bearded Ladies succeeded in spite of being female.

However, there is more than just altruism at play. Science itself benefits because diverse groups of problem solvers outperform high-ability problem solvers, particularly when faced with complex problems.[1] Having more voices, viewpoints, backgrounds, and ways of thinking results in better solutions!

Paleontology specifically is a science of complex problems. Paleontologists have to think in four dimensions: the three-dimensional world, plus time. And when paleontologists consider time, it is often at a magnitude well beyond that of a human lifetime or even the lifetime of the human species. Furthermore, fossil assemblages form an imperfect history of life on Earth, and so paleontological research can be like doing a puzzle where half the pieces are missing.

We call upon all of our readers to commit to building a workplace environment where everyone will thrive, and where, together, we can elevate the quality and quantity of the research performed. To provide support for the next generation of women in paleontology, The Bearded Lady Project: Currano Scholarship Fund was founded in 2017, in cooperation with the Paleontological Society. This fund is awarded as part of the society's annual Student Research Grant competition. Any student applying for a research grant who identifies as female will be considered for this scholarship. Each purchase of this book directly supports the scholarship fund.

For information, visit **paleosoc.org**.

---

1. Lu Hong and Scott E. Page, "Groups of Diverse Problem Solvers Can Outperform Groups of High-Ability Problem Solvers," *Proceedings of the National Academy of Sciences of the United States of America*, 101, no. 46 (2004): 16385-16389.

PLEDGE FOR EQUALITY
IN THE SCIENCES

I pledge…

To work toward an understanding and dismantling of gender bias;

To take an active stance against sexual assault and sexual harassment;

To support and advocate for equality and inclusivity for all, because our identities are intersectional and reflect a complex network of gender, race, sexuality, class, and ability;

To listen to and learn from the experiences of others and foster an environment in which all can learn and thrive: in higher education, my department, my lab, the classroom, the field, and beyond.

BIBLIOGRAPHY

American College of Obstetricians and Gynecologists. "Polycystic Ovary Syndrome (PCOS)." Accessed April 25, 2019. https://www.acog.org/Patients/FAQs/Polycystic-Ovary-Syndrome-PCOS.

"Annie Montague Alexander: Benefactress of UCMP." University of California Museum of Paleontology. Accessed October 16, 2019. https://ucmp.berkeley.edu/history/alexander.html.

Avery, Dan. "Trans Women Taught Me What a Denial Beard Is." NewNowNext, December 26, 2017. http://www.newnownext.com/denial-beards-transgender-women/12/2017/.

Burt, Ramsay. "The Trouble with the Male Dancer." In *Moving History/Dancing Cultures: A Dance History Reader*, ed. Ann Dils and Ann Cooper Albright, 44–55. Middletown, CT: Wesleyan University Press, 2001.

Butler, Judith. "Performative Acts and Gender Constitution: An Essay in Phenomenology and Feminist Theory." *Theatre Journal* 40, no. 4 (1988): 519–531.

Caplan-Bricker, Nora. "Is It Time to Change the Definition of 'Woman'?" Slate, September 29, 2017. https://slate.com/human-interest/2017/09/why-a-controversial-definition-of-the-word-woman-doesnt-necessarily-mean-the-dictionary-is-sexist.html.

Chocano, Carina. "When Gender Reveal Videos Go Spectacularly, Cathartically Wrong." *The New York Times Magazine*, August 1, 2019. https://www.nytimes.com/2019/08/01/magazine/gender-reveal-fail-videos.html.

Clark, Meredith. "5 Women with PCOS Explain Why They Choose to Celebrate Their Facial Hair." *Allure*, May 30, 2018. https://www.allure.com/story/women-with-pcos-facial-hair-beard-interviews.

Cohen, Phoebe A., Alycia Stigall, and Chad Topaz. "A Gender Analysis of the Paleontological Society: Trends, Gaps, and a Way Forward." Poster at North American Paleontological Convention, 2019.

Columbia Law School. "Kimberlé Crenshaw on Intersectionality, More Than Two Decades Later." Accessed October. 2, 2019. https://www.law.columbia.edu/pt-br/news/2017/06/kimberle-crenshaw-intersectionality.

Combahee River Collective. "Combahee River Collective Statement." In *Home Girls: A Black Feminist Anthology*, ed. Barbara Smith. New Brunswick, NJ: Rutgers University Press, 1983.

Diamond, Elin. "Introduction." In *Performance and Cultural Politics*, ed. Elin Diamond, 1–14. London and New York: Routledge, 1996.

Dutt, Kuheli, Danielle L. Pfaff, Ariel F. Bernstein, Joseph S. Dillard, and Caryn J. Block. "Gender Differences in Recommendation Letters for Postdoctoral Fellowships in Geoscience." *Nature Geoscience* 9 (2016): 805–808.

Dyer, Richard. *White*. London: Routledge, 1997.

Eagly, Alice H., and Antonio Mladinic. "Are People Prejudiced Against Women? Some Answers from Research on Attitudes, Gender Stereotypes, and Judgements of Competence." *European Review of Social Psychology* 5, no. 1 (1994): 1–35.

Ferla, Ruth La. "They're Mad As Hell." *The New York Times*, July 30, 2019. https://www.nytimes.com/2019/07/30/style/theyre-mad-as-hell.html.

Fisher, Donald W. "Memorial to Winifred Goldring 1888–1971." *Memorials of the Geological Society of America* 3 (1974): 96–107.

Flood, Allison. "Sexism Row Prompts Oxford Dictionaries to Review Language Used in Definitions." *The Guardian*, January 25, 2016. https://www.theguardian.com/books/2016/jan/25/oxford-dictionary-review-sexist-language-rabid-feminist-gender.

Garcia, Michelle. "Our Cover Star, Caster Semenya: The Athlete in the Fight for Her Life." *Out*, July 23, 2019. https://www.out.com/sports/2019/7/23/our-cover-star-caster-semenya-athlete-fight-her-life.

Ghosh, Pallab. "Cern Scientist: 'Physics Built by Men—Not by Invitation." BBC News, October 1, 2018. https://www.bbc.com/news/world-europe-45703700.

Gries, Robbie R. *Anomalies—Pioneering Women in Petroleum Geology: 1917–2017*. Lakewood, CO: Jewel Publishing LLC, 2017.

Halpern, Diane F., Camilla P. Benbow, David C. Geary, Ruben C. Gur, Janet Shibley Hyde, and Morton Ann Gernsbacher. "The Science of Sex Differences in Science and Mathematics." *Psychological Science in the Public Interest* 8, no. 1 (2007): 1–51.

Hay, Carol. "Who Counts as a Woman?" *The New York Times*, April 1, 2019. https://www.nytimes.com/2019/04/01/opinion/trans-women-feminism.html.

Hickman, Janell M. "Instagrammers Challenge Body and Facial Hair Stigma." *Teen Vogue*, March 28, 2017. https://www.teenvogue.com/story/girls-challenging-body-and-facial-hair-stigma.

Hong, Lu, and Scott E. Page. "Groups of Diverse Problem Solvers Can Outperform Groups of High-Ability Problem Solvers." *Proceedings of the National Academy of Sciences of the United States of America* 101, no. 46 (2004): 16385–16389.

hooks, bell. *Ain't I a Woman: Black Women and Feminism*. Boston: South End Press, 1981.

Hope, Allison. "She Invented the Gender Reveal Party. She Has Some Regrets." *Elle*, July 29, 2019. https://www.elle.com/culture/a28536376/gender-reveal-inventor-interview-jenna-karvunidis/.

Hyde, Janet S., and Marcia C. Linn. "Diversity—Gender Similarities in Mathematics and Science." *Science* 314, no. 5799 (2006): 599–600.

Kerr, Breena. "What Do Women Want?" *The New York Times*, March 14, 2019. https://www.nytimes.com/2019/03/14/style/womxn.html?searchResultPosition=1.

Key, Asia. "Woman, Womyn, Womxn: Students Learn About Intersectionality in Womanhood." *The Standard* (Missouri State University), March 27, 2017. http://www.the-standard.org/news/woman-womyn-womxn-students-learn-about-intersectionality-in-womanhood/article_c6644a10-1351-11e7-914d-3f1208464c1e.html.

Kinsey Institute. "Money, John." Accessed August 5, 2019. https://www.kinseyinstitute.org/about/profiles/john-money.php

Ko, Lisa. "Unwanted Sterilization and Eugenics Programs in the United States." *Independent Lens*, January 16, 2019. http://www.pbs.org/independentlens/blog/unwanted-sterilization-and-eugenics-programs-in-the-united-states/.

Lai, Calvin K., Kelly M. Hoffman, and Brian A. Nosek. "Reducing Implicit Prejudice." *Social and Personality Psychology Compass* 7 (2013): 315–330.

"LGBTQ+ Definitions." Trans Students Educational Resources. Accessed August 5, 2019. httsp://www.transstudent.org/definition.

Letzter, Rafi. "A Physicist Said Women's Brains Make Them Worse at Physics—Experts Say That's 'Laughable.'" LiveScience, October 2, 2018. https://www.livescience.com/63730-physicist-says-women-bad-at-physics.html.

MacNell, Lillian, Adam Driscoll, and Andrea N. Hunt. "What's in a Name: Exposing Gender Bias in Student Ratings of Teaching." *Innovative Higher Education* 40, no. 4 (2015): 291–303.

Madden, Kathryn. "The Dictionary Definition of 'Woman' Needs to Change." *Marie Claire*, July 5, 2019. https://www.marieclaire.com.au/dictionary-sexism-woman.

Merbruja, Luna. "3 Common Feminist Phrases That (Unintentionally) Marginalize Trans Women." Everyday Feminism, May 12, 2015. https://everydayfeminism.com/2015/05/feminist-phrases-marginalize-trans-women/.

Miller, Kevin, Deborah J. Vagins, Anne Hedgepeth, Kate Nielson, and Raina Nelson. *The Simple Truth About the Gender Pay Gap*. Washington, DC: American Association of University Women, 2018, 9–18. https://www.aauw.org/aauw_check/pdf_download/show_pdf.php?file=The_Simple_Truth.

Mitchell, Galen. "I Was a Bearded Lady—I Just Didn't Know It Yet." TransSubstantiation, May 16, 2017. https://transsubstantiation.com/i-was-a-bearded-lady-i-just-didnt-know-it-yet-1a1ba2b97c59.

Moraga, Cherríe, and Gloria Anzaldúa, eds. *This Bridge Called My Back: Writings by Radical Women of Color*. 2nd ed. Latham, NY: Kitchen Table, Women of Color Press, 1983.

Moss-Racusin, Corrine A., John F. Dovidio, Victoria L. Brescoll, Mark J. Graham, and Jo Handelsman. "Science Faculty's Subtle Gender Biases Favor Male Students." *Proceedings of the National Academy of Sciences of the United States of America* 109, no. 41 (2012): 16474–16479.

Nagle, Rebecca. "The Healing History of Two-Spirit, a Term That Gives LGBTQ Natives a Voice." The Huffington Post, June 30, 2018. https://www.huffpost.com/entry/two-spirit-identity_n_5b37cfbce4b007aa2f809af1.

"National Science Foundation and National Centre for Science and Engineering Statistics." Scientists and Engineers Statistical Data System Surveys: Survey Year 2013. Accessed October 3, 2019. https://ncsesdata.nsf.gov/us-workforce/2013/.

North, Anna. "'I am a Woman and I am Fast': What Caster Semenya's Story Says About Gender and Race in Sports." Vox, May 3, 2019. https://www.vox.com/identities/2019/5/3/18526723/caster-semenya-800-gender-race-intersex-athletes.

Oxford University Press. "OED Online." Accessed August 5, 2019. www.oed.com/view/Entry/77468.

Pember, Mary Anne. "'Two Spirit' Tradition Far from Ubiquitous Among Tribes." Rewire News, October 13, 2016. https://rewire.news/article/2016/10/13/two-spirit-tradition-far-ubiquitous-among-tribes/.

Peterson, Anne Helen. *Too Fat, Too Slutty, Too Loud: The Rise and Reign of the Unruly Woman*. New York: Plume, 2017.

Rae, Jetta. "'I Like to Consider Myself Genderful': Interview with Bearded Lady Little Bear Schwarz." Ravishly, January 7, 2015. https://www.ravishly.com/2015/01/07/interview-bearded-lady-little-bear-shcwarz.

Regan, Alex. "Should Women Be Spelt Womxn?" BBC News, October 10, 2018. https://www.bbc.com/news/uk-45810709.

Somerville, Siobhan. "Scientific Racism and the Emergence of the Homosexual Body." *Journal of the History of Sexuality* 5, no. 2 (1994): 243–266.

Steele, Claude M., and Joshua Aronson. "Stereotype Threat and the Intellectual Test Performance of African-Americans." *Journal of Personality and Social Psychology* 69, no. 5 (1995): 797–811.

Symonds, Matthew R. E., Neil J. Gemmell, Tamsin Braisher, and Kylie Gorringe. "Gender Differences in Publication Output: Towards an Unbiased Metric of Research Performance." *Plos One* 1, no. 1 (2006): 5.

Taylor, Diana. *The Archive and the Repertoire: Performing Cultural Memory in the Americas*. Durham, NC: Duke University Press, 2003, 3.

Terry, Jennifer. "Lesbians Under the Medical Gaze: Scientists Search for Remarkable Differences." *The Journal of Sex Research* 27, no. 3 (1990): 317–339.

U.S. Department of Health and Human Services, Office of Women's Health. "Polycystic Ovary Syndrome." Accessed April 25, 2019. https://www.womenshealth.gov/a-z-topics/polycystic-ovary-syndrome#17.

Walker, Leslie. "How to Edit Gender Identity Status on Facebook." Lifewire, September 28, 2019. https://www.lifewire.com/edit-gender-identity-status-on-facebook-2654421.

Weiss, Suzannah. "Bearded Lady Little Bear Schwarz Makes Facial Hair 'Intensely Feminine.'" Vice, January 26, 2016. https://www.vice.com/en_us/article/4xkznq/bearded-lady-little-bear-schwarz-makes-facial-hair-intensely-feminine.

Wilson, Carolyn. *Status of the Geoscience Workforce*. Alexandria, VA: American Geosciences Institute, 2018.

Wynn, Natalie. "Transtrenders." ContraPoints, July 1, 2019. https://www.youtube.com/watch?v=EdvM_pRfuFM.

Young, Stephanie L. "Running Like a Man, Sitting Like a Girl: Visual Enthymeme and the Case of Caster Semenya." *Women's Studies in Communication* 38 (2015): 331–350.

"Yusra: Expert Excavator of Mount Carmel." Trowelblazers. Accessed October 16, 2019. https://trowelblazers.com/yusra-expert-excavator-of-mount-carmel/.

ACKNOWLEDGMENTS

The Bearded Lady Project would like to acknowledge all the women who not only participated in this project but—first and foremost—supported us with their friendship and love. Additionally, we would like to thank the following people:

Leslie Ching

Diane Currano

John Currano

Louis and Florence DeBiasio

Joan Greene

Jack Jamieson

Peter Jamieson

Brian Marsh

Ljuba Marsh

Gregg Randolph

Dr. Sarah Siff

Jim Vance

Marsha White

The Bearded Lady Project was made possible by funding and donations from:

The National Science Foundation

American Frame

The Paleontological Society

The University of Wyoming

THE PHOTOGRAPHERS

Kelsey Vance (portrait photographer) is a fine art photographer based in Ohio. Her photographs are an exploration of the environment she finds herself in as she seeks to understand her surroundings and establish a sense of belonging. Her images are often small observations of quiet contemplation. The portraits for *The Bearded Lady Project* follow similar themes as she attempts to create the visual presence of women that has been absent throughout the history of science. Kelsey earned a B.F.A. in photography from Maine College of Art and an M.F.A. in photography from Arizona State University.

Draper White (cinematographer and portrait photographer, ReBecca Hunt-Foster and Kelli Trujillo) is an award-winning photographer, retoucher, and cinematographer based in Basalt, Colorado, since 2015. He is one of the lucky few who have truly found their passion in their profession. Draper's commercial work takes him around the globe to work with international brands in the architecture and outdoor adventure industries.

Laura Dempsy (portrait photographer, Ashley Hall) combines photography, graphic design, video, illustration, and creative marketing to help businesses and individuals tell their stories. Passionate about lifelong learning, she also can be found bicycling, swimming, or enjoying the beauty of nature.